PHARMACEUTICAL CHEMISTRY-I

PHARMACEUTICAL CHEMISTRY-I

(for D. Pharm. Part-I Students as per P.C.I. Syllabus, E.R. 1991)

Dr. Anil Bhandari
Senior Lecturer, Dept. of Pharmacy
L.M. College of Science & Technology,
Jodhpur (Rajasthan)

Dr. G.K. Singh
Lecturer, Dept. of Pharmacy,
L.M. College of Science & Technology
Jodhpur (Rajasthan)

CBSPD

CBS Publishers & Distributors Pvt Ltd

New Delhi • Bengaluru • Chennai • Kochi • Kolkata • Lucknow • Mumbai
Hyderabad • Jharkhand • Nagpur • Patna • Pune • Uttarakhand

PHARMACEUTICAL CHEMISTRY-I

ISBN: 978-81-239-0630-0

First Edition: 1998
Reprint: 2001, 2003, 2004, 2005, 2006, 2007, 2008, 2010, 2012, 2015, 2016, 2018, 2019, 2020, 2023

Published by **Satish Kumar Jain** and produced by **Varun Jain** for
CBS Publishers & Distributors Pvt Ltd
4819/XI Prahlad Street, 24 Ansari Road, Daryaganj, New Delhi 110 002, India.
Ph: 011-23289259, 23266861, 23266867
Fax: 011-23243014
Website: www.cbspd.com
e-mail: delhi@cbspd.com;

Corporate Office: 204 FIE, Industrial Area, Patparganj, Delhi 110 092
Ph: 011-4934 4934 Fax: 011-4934 4935
e-mail: publishing@cbspd.com; publicity@cbspd.com

Branches

- **Bengaluru:** Seema House 2975, 17th Cross, KR Road, Banasankari 2nd Stage, Bengaluru 560 070, Karnataka, India
  Ph: +91-80-26771678/79 Fax: +91-80-26771680 e-mail: bangalore@cbspd.com
- **Chennai:** 7, Subbaraya Street, Shenoy Nagar, Chennai 600 030, Tamil Nadu, India
  Ph: +91-44-26680620, 26681266 Fax: +91-44-42032115 e-mail: chennai@cbspd.com
- **Kochi:** 42/1325, 1326, Power House Road, Opp KSEB, Power House, Ernakulum Kochi 682 018, Kerala, India
  Ph: +91-484-4059061-65,67 Fax: +91-484-4059065 e-mail: kochi@cbspd.com
- **Kolkata:** 147, Hind Ceramics Compound, 1st Floor, Nilgunj Road, Belghoria, Kolkata-700056, West Bengal, India
  Ph: +033-25633055, 033-25633056 e-mail: kolkata@cbspd.com
- **Lucknow:** Basement, Khushnuma Complex, 7 Meerabai Marg (Behind Jawahar Bhawan),Lucknow-226001, UP, India
  Ph: +0522-4000032 e-mail: tiwari.lucknow@cbspd.com
- **Mumbai:** PWD Shed, Gala no 25/26, Ramchandra Bhatt Marg, Next to JJ Hospital Gate no. 2, Opp. Union Bank of India, Noorbaug, Mumbai-400009, Maharashtra, India
  Ph: 022-66661880/89 e-mail: mumbai@cbspd.com

Representatives

- Hyderabad 0-9885175004
- Patna 0-9334159340
- Jharkhand 0-9811541605
- Pune 0-9923910676
- Nagpur 0-9421945513
- Uttarakhand 0-9716462459

Printed at Neekunj Print Process, Haryana, India

Preface

This book has primarily been written with a view to meeting the needs and interest of the Diploma in Pharmacy Part-I student to cover the syllabus of Pharmaceutical Chemistry I, framed by Pharmacy Council of India (P.C.I.) as per E.R. 1991.

This book has been divided into six chapters to suit the P.C.I. syllabus. The book contains important tests for purity given in Indian Pharmacopoeia 1996, assays, preparations and chemical properties of certain compounds. The appendix covers certain definitions and medical terms which may be of helpful to the students.

We hope that this book shall be useful for students and teachers of diploma in Pharmacy as well as B.Pharm. students. We welcome suggestions and comments for improvement of the book.

We express our sincere thanks to Prof. M.M. Mathur for first reading and to Mr. S.K. Jain, CBS Publishers and Distributors, Delhi for publishing this book.

Anil Bhandari
G.K. Singh

Syllabus

1. General discussion on the following inorganic compounds including important, physical and chemical properties, medicinal and pharmaceutical uses, storage conditions and chemical incompatibility.

   (a) **Acids, Bases and Buffers** : Boric acid, Hydrochloric acid, Strong ammonium hydroxide, Calcium hydroxide, Sodium hydroxide and Official buffers.

   (b) **Antioxidants** : Hypophosphorous acid, Sulphur dioxide, Sodium bisulphite, Sodium meta-bisulphite, Sodium thiosulphate, Nitrogen and Sodium Nitrite.

   (c) **Gastrointestinal Agents**

   (i) **Acidifying Agents** : Dilute hydrochloric acid.

   (ii) **Antacids** : Sodium bicarbonate, Aluminium hydroxide gel, Aluminium phosphate, Calcium carbonate, Magnesium carbonate, Magnesium trisilicate, Magnesium oxide, Combinations of antacid preparations.

   (iii) **Protectives and Adsorbents** : Bismuth subcarbonate and Kaolin.

   (iv) **Saline Cathartics** : Sodium potassium tartrate and Magnesium sulphate.

   (d) **Topical Agents**

   (i) **Protectives** : Talc, zinc oxide, Calamine, Zinc stearate, Titanium dioxide, Silicone polymers.

   (ii) **Antimicrobials and Astringents** : Hydrogen peroxide, Potassium permanganate, Chlorinated lime, Iodine, Solutions of iodine, Povidone iodine, Boric acid, Borax, Silver nitrate, Mild silver protein, Mercury, Yellow mercuric oxide, Ammoniated mercury.

   (iii) **Sulphur and its Compounds** : Sublimed sulphur, Precipitated sulphur, Selenium sulphide.

   (d) **Astringents** : Alum and zinc Sulphate.

   (e) **Dental Products** : Sodium fluoride, Stannous fluoride, Calcium carbonate, Sodium meta phosphate, Dicalcium phosphate, Strontium chloride, Zinc chloride.

(f) **Inhalants :** Oxygen, Carbon dioxide, Nitrous oxide.

(g) **Respiratory Stimulants :** Ammonium carbonate.

(h) **Expectorants and Emetics :** Ammonium chloride, Potassium iodide, Antimony potassium tartrate.

(i) **Antidotes :** Sodium nitrite.

2. Major Intra and Extracellular Electrolytes

(a) **Electrolytes used for Replacement Therapy :** Sodium chloride and its preparations; Potassium chloride and its preparations.

(b) **Physiological Acid-base Balance and Electrolytes used :** Sodium acetate, Potasium acetate, Sodium bicarbonate injection, Sodium citrate, Potassium citrate, Sodium lactate injection, Ammonium chloride and its injection.

(c) Combination of oral electrolytes powders and solutions.

3. Inorganic Official Compounds of Iron, Iodine and Calcium, Ferrous Sulphate and Calcium Gluconate.

4. **Radiopharmaceuticals and Contrast Media :** Radioactivity, Alpha, Beta and Gamma Radiations, Biological Effects of Radiation, Measurement of Radioactivity G.M. Counter - Radio isotopes their Uses, Storage and Precautions with Special Reference to the Official Preparations.

Radio-opaque Contrast Media-Barium sulphate.

5 **Quality Control of Drugs and Pharmaceuticals :** Importance of Quality Control, Significant Errors, Methods used for Quality Control, Sources of Impurities in Pharmaceuticals. Limit Tests for Arsenic, Chloride, Sulphate, Iron and Heavy Metals.

6. Identification Tests for Cations and Anions as per Indian Pharmacopoeia.

Contents

1

Quality Control of Drug and Pharmaceuticals

Importance of quality control

Quality control is vital for substances, which are used internally and / or externally on the living organism. These substances are used either to alter the physiological action or to remove unwanted microrganisms or to enhance the appearance, i.e., to act as a drug or cosmetic. These agents should be superior in quality and must confirm to the established quality standards. Every country determines specific quality standards for each drug and cosmetic. These official parameters are published in a book called Pharmacopoeia e.g., Indian official book is called Indian Pharmacopoeia (I.P.); Britain's official book is called British Pharmacopoeia (B.P.); similarly, Japanese-Pharmacopoeia (J.P.); United State Pharmacopoeia (U.S.P.), etc.

Indian Pharmacopoeia : The word pharmacopoeia has been derived from Greek word "Pharmakon", meaning drug or medicine and "poieo", to make. In India B.P. was used as official standard book before independence, but in 1946 government published Indian pharmacopoeial list in addition to British pharmacopoeia, and an Indian pharmacopoeial committee was constituted in 1948. This was reconstituted in 1954 under the chairmanship of Dr. B.N. Ghosh, Professor of Pharmacology, R.G. Kar Medical College, Calcutta. The first edition of pharmacopoeia was published in 1955. Unfortunately Dr. B.N. Ghosh died in 1958 and Dr. B. Mukerjee was appointed as Chairman of the committee for the publication of IInd edition. The second edition of Indian pharmacopoeia was published in 1966. Its supplement was published in 1975. The third pharmacopoeial committee was appointed under the chairmanship of Dr. Nityanand, Director, C.D.R.I., Lucknow and IIIrd edition was published in 1985. The fourth Indian pharmacopoeial committee was set up under the chairmanship of Dr. Nitryanand. The IVth edition was published in 1996. The Indian pharmacopoeia's enlists various inorganic and organic chemicals and their official preparations, standards, physical characteristics, solubility, identification tests, purity tests, assay, storage, category and doses of various substances. A list of appendix is given to help in performing various procedures.

Accuracy : Accuracy is defined as the concordance between procedural value and true or most probable value. Accuracy may be attained with absolute methods like weighing of pure substance, purification or recrystallization of primary standard substances etc., and comparative methods involved in secondary standards and various analytical methods.

Precision : Precision is defined as "concordance of a series of measurements of the same quality". Accuracy expresses the correctness of a measurement while precision shows reproducibility of a measurement.

Errors : There are two types of errors-(i) Determinate or constant errors (which can be avoided or magnitude of which can be determined) (ii) Indeterminate or accidental errors (errors made on successive measurement in nearly identical conditions).

Minimization of errors : The errors can change accuracy of the result. These errors can be reduced by following methods.

1. Calibration of apparatus
2. Running a control determination
3. Running of parallel determination
4. Addition of standard substances
5. Amplification of method
6. Isotopic dilution.

These methods help in determining possible determinate or indeterminate errors, so that these errors can be rectified or possible preventive measures can be taken.

Significant figures : The computation of experimental results can be done by following methods.

1. The mean (average) of all the results is calculated and the difference of average value and each result will give deviation from mean value. It can be converted into percentage mean deviation.

$$\text{Percentage mean deviation} = \frac{\text{deviation}}{\text{mean value}} \times 100$$

2. The relative deviation is calculated by assuming one value as accurate value and difference between that accurate value and individual value is relative deviation.
3. Model deviation : All the values are classified in various batches and a batch containing maximum values is called model batch. Average of model batch is called model value. If a batch is not formed, the values which are repeated maximally or values which are concordant, are model values. The difference between model value and individual value is called model deviation.

All these values are used to find out standard deviation and coefficient of variation.

Standard deviation (S.D.) $= \sqrt{(X-\bar{X})^2/n-1}$

Coefficient of variation (CV) $= 100 \text{ S.D. } \sqrt{X}$

Where $\bar{X}$ = mean or model value

n = number of values

$(X-\bar{X})$ = mean or model deviation

Quality Control

The quality control of various pharmaceutical drugs and dosage form should be in accordance with official books. The production should be carried out with respective schedule M requirement, adhering to good manufacturing practices (G.M.P.). The quality control and analytical control, i.e., description, solubilities, standards, identification, test for purity and assay procedures should be according to Pharmacopoeial monograph. In fixing these standards, it is necessary to have the knowledge about sources of impurities, their effects and possibilities of their removal.

Source of impurities in pharmaceutical substances

Chemical purity means being free from foreign substances. It is very costly to manufacture absolutely pure chemical substances. If a product with small impurity is comparatively cheap and still safe in medicinal use, it can be used for formulation of dosages form. The pharmacopoeia prescribes various tests for purity for each substance and also fixes limits of tolerance for certain impurities like chlorides, sulphates, iron, heavy metals, arsenic and lead. These impurities are commonly present in pharmacopoeial substances in very small amount. Arsenic and lead are cumulative (these are neither excreted nor catabolised in the body) and, therefore they are accumulate in the body and produce toxicity.

There are following sources of impurities in pharmaceutical substances.

(1) **Raw materials :** If impurities are present in the raw materials (ores, metals etc.). These may pass on during the manufacturing process to the final products. Thus the final compound may be contaminated with these impurities. e.g.

I. Zinc sulphate is prepared by either zinc oxide or zinc metal with sulphuric acid.

$$ZnO + H_2SO_4 \longrightarrow ZnSO_4 + H_2O$$

$$Zn + H_2SO_4 \longrightarrow ZnSO_4 + H_2$$

Zinc metal and zinc oxide contain Al, Cu, Mg, Mn, Ni, As, and Fe as impurities and these impurities are likely to be carried in the manufacturing process in appreciable amounts to the final product as $ZnSO_4$ $7H_2O$.

II. Copper Sulphate is prepared by the reaction of copper turnings with sulphuric acid.

$$Cu + 2H_2SO_4 \longrightarrow CuSO_4 + 2H_2O + SO_2$$

Copper turnings may contain Fe and As impurities. If these impurities are present in appreciable amount in raw material then the final product $CuSO_4.5H_2O$ may be contaminated.

(2) **Reagents used in manufacturing process :** The reagents which are used in the manufacturing process; if not removed completely by washing, may still be present in the final product. e.g.

I. When calcium chloride is roasted with sodium carbonate, precipitated calcium carbonate is formed.

$$\underset{\text{(Soluble)}}{CaCl_2} + \underset{\text{(Soluble)}}{Na_2CO_3} \longrightarrow \underset{\text{(ppt.)}}{CaCO_3} + \underset{\text{(Soluble)}}{2NaCl}$$

The precipitate of calcium carbonate is washed to remove excess of sodium carbonate and calcium chloride. If the precipitate is not properly washed, it may persist as an impurity. The pharmacopoeia prescribes limits of tolerance for soluble alkali (Na_2CO_3) and chlorides as impurities.

II. Ammoniated mercury is prepared by the action of mercuric chloride with dilute ammonia solution.

$$\underset{\text{(Soluble)}}{HgCl_2} + \underset{\text{(Soluble)}}{NH_4OH} \longrightarrow \underset{\text{(ppt.)}}{NH_2HgCl} + \underset{\text{(Soluble)}}{NH_4Cl + 2H_2O}$$

The NH_2HgCl is completely washed to remove dilute ammonia. If NH_2HgCl is not completely washed, the final product may contain ammonia, as impurity.

(3) **Intermediate products in the manufacturing process :** Sometimes an intermediate substance produced during the manufacturing process may remain in the final product. e.g. Potassium iodide is prepared by the reaction of potassium hydroxide and iodine.

$$6KOH + 3I_2 \longrightarrow 5KI + KIO_3 + 3H_2O$$

The resulting solution is evaporated to dryness and the residue is heated with charcoal.

$$KIO_3 + 3C \longrightarrow KI + 3CO$$

Potassium iodate so formed is an intermediate product and if it is not converted into potassium iodide, it may be present as impurity in the final product, and hence Pharmacopoeia describes a test for iodate.

(4) **Defects in manufacturing process :** Defects like incompleteness of reaction, imperfect mixing etc. may result in the production of impure compounds. e.g.

I. Zinc-oxide is produced by heating metallic zinc in a current of air.

$$2Zn + O_2 \longrightarrow 2ZnO$$

If Zinc metal is not completely converted into Zinc oxide, small amount of Zinc metal may be present in the final product as impurity. Therefore, Indian pharmacopoeia prescribes test for Zinc metal.

(5) **Solvents :** Water is the cheapest solvent and is used universally. Water may give impurities like chloride, sulphate, bicarbonate, magnesium, calcium, etc. To remove these impurities from water, either distilled water or deionized water should be used.

(6) **Action of solvent and reagent on reacting vessel :** Some reagents and solvents may react with the container, in which they are stored or processed. e.g.

Strong acid leaches out alkali from borosilicate glass; Copper and Zinc vessels react with slightly acidic substances, etc. Hence a great care must be taken in selection of suitable vessel for manufacturing processes or for the purpose of storage.

(7) **Atmospheric contamination :** Atmospheric air may contain CO_2, SO_2, H_2S, or dust particles of alumina Al_2O_3, Carbon, Silica glass particles, plastic fragments, etc, depending on the areas where the industries are located e.g.

Solution of sodium hydroxide absorbs atmospheric carbon dioxide to form sodium carbonate and bicarbonate.

(8) **Faulty Storage :** The improper storage of substances may change their physicochemical properties. e.g.

When hydrogen peroxide and silver nitrate are stored in glass bottle, these solutions get deteriorated; When ferrous sulphate is exposed to moist air, ferric sulphate is formed; Potassium iodide and ferric ammomium sulphate liquidize when exposed to moist air.

(9) **Adulteration :** For the purpose of increase in bulk, increase in weight or decrease in cost, one may be tempted to mix some bulky agent. Higher molecular weight substance or cheaper agents of similar properties are incorporated to increase the weight and bulk of drug, so as to increase profitability. Adulteration beyond permissible impurities is punishable under, drug and cosmetic act 1940.

Permissible impurities

It is impossible to obtain 100% pure substance or in otherwords, it is difficult to remove all impurities. Hence the limit of maximum impurity which may be allowed to be present in best quality of substance is fixed by considering following parameters.

1. The impurities present in the substance is determined. Maximum allowable limits are fixed as per their tolerance level.
2. If the process of removal of impurities are not very difficult, maximum impurities should be removed.
3. The cost effectiveness of the process which means that the cost of product should be reasonable. e.g.

   Commercial sodium Chloride is Rs. 4.50 per kg, while 99.5% pure pharmaceutical grade (I.P.) NaCl is around Rs. 40 per kg. but if we prepare impurity free NaCl, it may cost Rs. 400 per kg. approximately.
4. The tolerance level of toxic substances like arsenic, lead, other heavy metals, etc. should be determined and in any case these elements level, should be less than tolerance level.

Test for purity or test for impurity

The common impurities found in most substances are chloride, sulphate, iron, lead, heavy metals and arsenic. The permissible limit of these impurities does not vary much from compound to compound and procedure is also similar. Test for these impurities is called limit test. Limit test is a qualitative as well as semi quantitative test. Except arsenic all these limit tests are performed in a pair cylinders having, uniform diameter, i.e, nessler cylinders as per the specification of pharmacopoeia. The limit test of arsenic is performed in a specially designed glass apparatus called "Gutzeit apparatus" in the honour of scientist "Gutzeit". The test for other impurities present in the substance is called "Test for purity"

Limit test for chlorides

Test	Standard	Reasons
1. Dissolve specified quantity of substance as given in monograph or prepare a solution as directed in individual monograph.	1. Place 1 ml. of a 0.0585 % W/v solution of sodium chloride in Nessler cylinder labelled as "standard". Add about 9 ml of distilled water and mix.	1. The aqueous solution will leach out all the chloride ion present in the sample or test so that they can be ready to react with silver nitrate and opalescence can be compared.
2. Add 10 ml of dilute nitric acid.	2. Add 10 ml of dilute nitric acid.	2. The nitric acid prevents the precipitation of other radicals like phosphate, sulphate, etc. This produces common ion effect with silver nitrate and dissociation of silver ion is supressed $HNO_3 \longrightarrow H^+ + NO_3$
3 Dilute to 50 ml. with chloride free water.	3. Dilute to 50 ml with chloride free water	3. Limit test is a comparison hence equal volume can be compared easily.
4. Add 1 ml of 0.1 M $AgNO_3$, stir and allow to stand for 5 minutes in dark.	4. Add 1 ml 0.1 M $AgNO_3$, stir and allow to stand for 5 minutes in dark.	4. The Ag^+ ion react with Cl^- ion to form opalescence of silver chloride. $Ag^+ + Cl^- \dashrightarrow AgCl$

If opalescence produced in standard is more than that in the test, the sample complies limit test of chloride.

Limit test for sulphate (According to I.P. 1985)

Test	Standard	Reasons
1. Dissolve specified quantity of substance as given in monograph or prepare a solution as directed in individual monograph.	1. Place 1 ml of 0.1089 W/v solution of potassium sulphate in Nessler cylinder labelled as "standard" Add about 9 ml of distilled water	1. The aqueous solution will leach out all the sulphate ion present in the sample or test.
2. Add 2 ml of dilute hydrochloric acid	2. Add 2 ml of dilute hydrochloric acid.	2. Hydrochloric acid suppreses dissociation of $BaCl_2$ and less barium ions are form which prevents the precipitation of other acid radicals like Oxalate, phosphate etc. $HCl \longrightarrow H^+ + Cl^-$ $BaCl_2 \longrightarrow Ba^{+2} + 2Cl^-$
3. Dilute to 45 ml with distilled water (free from sulphate)	3. Dilute to 45 ml with distilled water (free from sulphate)	3. Distill water is added to make equal volumes of test and standard.
4. Add 5 ml of $BaSO_4$ reagent	4. Add 5 ml of $BaSO_4$ reagent	4. (a) Barium sulphate reagent contains small amount of K_2SO_4 (0.905 mg in 100 ml) which increases the sensitivity of limit-test (b) $BaSO_4$ reagent contain small quantity of alcohol, which prevents the super saturation of $BaSO_4$- turbidity to keep it in precipitated from. (c) The major ingredient of this reagent is $BaCl_2$, which precipitates (turbidity) sulphate ions and turbidity can be compared. $BaCl_2 + SO_4^{-2} \longrightarrow BaSO_4 + 2Cl^-$

If turbidity produced in standard is more than that in test, the sample complies limit test of sulphate as per I.P. 1985.

Limit test for sulphate (As per I.P. 1996)

Test	*Standard*	*Reasons*
1. Take 1 ml of 25 % W/v soln. of $BaCl_2$ in nessler cylinder.	1. Take 1 ml of 25 % W/v soln. of $BaCl_2$ in Nessler cylinder	1. It is a principle precipitating agent. The Barium ions reacts with sulphate to form opalescence of barium sulphate. $BaCl_2 + SO_4^{-2} \longrightarrow BaSO_4 + 2\,Cl^-$
2. Add 1.5 ml of ethanolic sulphate standard solution (10PPM of SO_4^{-2})	2. Add 1.5 ml. of **ethanolic** sulphate standard sulution (10 PPM of SO_4^{-2})	2. This ethanolic solution of potassium sulphate increases sensitivity of the test and alcohols prevents the supersaturation of barium sulphate precipitate.
3. Add 15 ml of solution prepared as directed in individual monograph.	3. Add 15 ml of standard sulphate solution (10 PPM or 10 PPM sulphate).	3. It gives sulphate ion present in the sample of sulphate ion in standard.
4. Add 0.15 ml of 5M acetic acid.	4. Add 0.15 ml of 5M with acetic acid.	4. The acetic acid prevents the precipitation of various anions such as borate, oxalate, phosphates, etc.
5. Makeup the volume with sulphate free water upto 50 ml	5. Make up the volume with sulphate free water upto 50 ml.	5. Equal volume opalescence are easily compared by viewing in black background.

If Opalescence produced in the standard is more than that in the test, the sample complies limit test of sulphate as per I.P. 1996.

Limit Test For Iron

Test	*Standard*	*Reasons*
1. Dissolve specified quantity of substance as given in monograph or prepare a solution as directed in individual monogrph.	1. Place 2 ml of the standard iron solution (equal to 0.04 mg Fe) in Nessler cylinder and add 40 ml of distilled water.	1. The aqueous solution will leach out all the iron ion present in the sample or test.
2. Add 2 ml of 20 % solution of iron free citric acid (FeT.)	2. Add 2ml of 20% solution of iron free citric acid (FeT)	2. Citric acid prevent the precipitation of Iron with ammonia and keeps iron in solu tion even in presence of ammonia by forming complex with iron as iron citrate. $2Fe+10NH_3 \longrightarrow 2Fe(NH_2)_5 +5H_2$ $C_6H_8O_4 + Fe \longrightarrow Fe(C_6H_6O_4)_2 + 2H_2$
3. Add 0.1 ml of thio-glycollic acid (FeT)	3. Add 0.1 ml of thio-glycollic acid (FeT)	3. Thioglycollic acid converts ferric ions to ferrous ions. $2HSCH_2COOH+Fe^{+3} \longrightarrow Fe^{+2}+2H^{+}$. $HSCH_2COO^{-2}$ Thio-glycollic acid reacts with ferrous ion to form ferrous thioglycolate. $HSCH_2COOH+Fe^{+2} \longrightarrow (HSCH_2COO)_2Fe + 2H^+$
4. Make the solution alkaline with iron free ammonia solution (FeT)	4. Make the solution alkaline with iron free ammonia solution (FeT)	4. The ferrous thioglycolate formed is colourless in acidic medium but in alkaline medium it gives purple colour.
5. Dilute to 50 ml mark with distilled water.	5. Dilute to 50 ml. mark with distilled water.	5. Equal volume of test and standard are compared easily.

If the intensity of purple colour in standard is more than that in the test, the sample complies limit test of iron as per I.P. 1996.

FeT :-- Tested for iron free.

Limit test for lead

Lead is the most undesirable and common impurity found in medicinal compounds. Common source of lead impurities are :

1. Drugs prepared from sulphuric acid.
2. Defective storage of medicinal compounds, in ordinary lead glass bottles and
3. Containers are opened repeatedly in moist conditions or kept without closure.

Limit test (I.P. Method)

Principle : In I.P. 1996 ; the limit test for lead is based on the reaction between lead and dithizone (diphenylthiocarbazone) which results in the formation of lead dithizonate. In this test the Nessler cylinders are not used. The test is performed by extraction with the help of separating funnel.

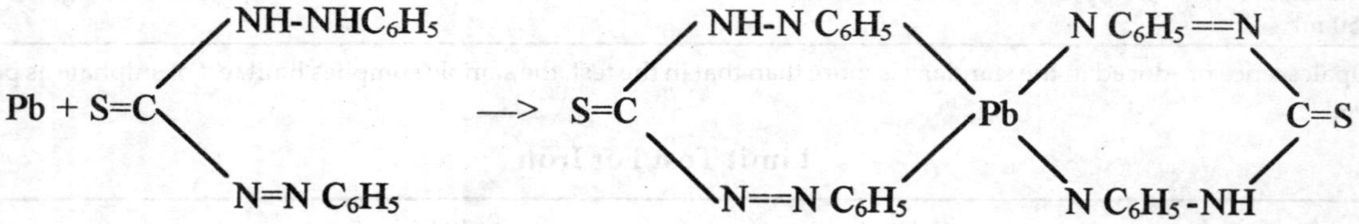

Dithizone Lead dithizonate

Dithizone is soluble in chloroform and gives a green coloured solution. It is called dithizone chloroform extraction solution and is used for quantitative extraction of lead. The lead dithizonate (Final product) in chloroform is red in colour. Mixture of these two (green and red coloured solutions) appears as violet in colour.

Intensity of violet colour in the test sample should not be more than that of the standard solution produced in the chloroform medium.

Procedure : Transfer the prepared sample to a separator, add 6 ml of ammonium citrate solution Sp and 2 ml of hydroxylamine hydrochloride solution Sp and 2 drops of phenol red solution and prepare the solution alkaline by the addition of strong ammonia solution, cool and add 2ml of potassium cyanide solution. Immediately extract the solution with several quantities of 5 ml of dithizone extract, draining off each extract into another separating funnel, until the dithizone extraction solution retain its green colour. Shake the combined dithizone solution for 30 sec. with 30 ml of 1% V/V solution of nitric acid and discard the chloroform layer. Mix the acid solution with 5 ml of dithizone standard solution and shake for 30 seconds. The colour of chloroform extract is not more intense than that obtained by treating in the same manner a solution of lead standard (1 ppm Pb).

Limit Test for Heavy Metals

The Indian pharmacopoeia (1996) prescribes limit test of heavy metals for some drugs. The heavy metals like cobalt, tin, manganese, bismuth, lead, antimony etc., are present in the pharmacopoeial substances. The Indian pharmacopoeia (1996) prescribes methods for the limit test of heavy metals.

1. **Method 'A' :**

Principle : This method is used for the colourless substance. In acidic medium when hydrogen sulphide is reacted with heavy metals, the metal sulphites are produced. These are distributed in colloidal state and produce brown colour. The 'test' colour is compared with "standard".

$$\underset{\text{Heavy metal}}{M + H_2S} \xrightarrow[\text{medium}]{\text{acidic}} \underset{\text{(Brown colouration)}}{MS + H_2}$$

Standard Solution : Take 1.0 ml of lead standard solution in 50 ml Nessler cylinder and dilute with water to 25 ml. Adjust with dilute ammonia solution or dilute acetic acid so that pH is maintained between 3.0 and 4.0, then dilute with water to 35 ml and mix.

Test Solution : In 50 ml of Nessler cylinder take 25 ml of the solution or specified quantity of substance being examined, prepared for the test as directed in the individual monograph in sufficient water to produce 25 ml. Adjust with dilute acetic acid or dilute ammonia solution so that pH is maintained between 3.0 to 4.0, dilute with water to 35 ml and mix.

Procedure : In both Nessler cylinders add 10 ml of freshly prepared H_2S solution, mix, dilute to 50 ml with water, allow to stand for 5 minutes and see downward over a white surface.

The colour produced in the "Test Solution" should not be more intense than that produced in the "Standard solution".

2. **Method 'B'**

Principle : Method 'B' is used for the coloured substance. The coloured substances are converted into colourless before preparing their solution with special treatment (Sulphuric acid, ignition, nitric acid, digested with hydrochloric acid, and finally react with water). This method is based on the reaction between H_2S and heavy metals in an acidic medium and metal sulphide is produced.

$$\text{Heavy Metal } H_2S \xrightarrow[\text{Medium}]{\text{Acidic}} \underset{\text{(Brown Colouration)}}{\text{Sulphides of heavy metals}}$$

Standard Solution : Take 1.0 ml of standard lead solution in 50 ml of Nessler cylinder and dilute with water to 25 ml. Adjust with dilute ammonia solution or dilute acetic acid, so that pH is maintained between 3.0 and 4.0. Then dilute with water to 35 ml and mix.

Test Solution : Weigh in a silica crucible, the quantity of the substance as specified in the individual monograph. Add sulphuric acid to wet the sample, ignite carefully at a low temp.

until thoroughly charred. Add 2 ml of nitric acid to charred mass and 5 drops of sulphuric acid and heat until white fumes are evolved. Ignite at 500° to 600°C, until the carbcn is completely burnt off. Cool and add 4 ml of hydrochloric acid; digest on a water bath for 15 minutes, and then slowly evaporate to dryness on a water bath. Moisten the residue with 1 drop hydrochloric acid, add 10 ml of hot water and digest for 2 minutes. Mix ammonia solution in sufficient quantity to make alkaline add 25 ml of water and adjust pH between 3.0 and 4.0 with dilute acetic acid. Filter, if required; wash the crucible and the filter with 10ml of water, dilute with water to 35 ml and mix.

Procedure : Add 10 ml of freshly prepare H_2S solution in both the Nessler cylinders, mix, dilute to 50 ml with water, allow to stand for 5 minutes. The colour produced in the "test" is compared with that of the "standard solution".

3. Method 'C'

Principle : Method C is used for those substances which produce clear and colourless solution with sodium hydroxide. When heavy metals react with sodium sulphide in alkaline medium, heavy metal sulphides are produced.

$$\text{Heavy Metal} + Na_2S \xrightarrow[\text{Medium}]{\text{Alkaline}} \text{Sulphides of heavy metals (Dark Colouration)}$$

Standard Solution : Pipette out 1.0 ml of lead solution into 50 ml Nessler cylinder. Add 5 ml of dilute sodium hydroxide solution and dilute with water to 50 ml and mix.

Test Solution : Place 25 ml of the solution prepared for the test as directed in the individual monograph into a 50 ml Nessler cylinder, or dissolve the specified quantity of the substance being examined in a mixture of 20 ml of water and 5 ml of dilute sodium hydroxide solution. Dilute with water to 50 ml and mix.

Procedure : Add 5 drops of sodium sulphide solution in both Nessler cylinders (standard and test), mix and allow to stand for 5 minutes. The colour produced in the 'test' is compared with that of the "Standard Solution".

4. Method 'D'

Standard Solution : Pipette 10.0 ml of lead standard solution (1 ppm Pb) in to a small Nessler cylinder.

Test Solution : Prepare sample solution as directed in the individual monograph in small Nessler cylinder.

Procedure : Add 2.0 ml of the 'test solution' in the cylinder containing the "standard solution" and mix. Add 2 ml of acetate buffer,(pH 3.5),in each cylinder and mix. Add 1.2 ml of thioacetamide reagent. Allow to stand for 2 minutes and view downwards over a white surface. The colour produced in the 'test solution' and "standard solution" are compared.

Limit Test for Arsenic

Principle : Arsenic produces cumulative toxicity, and hence presence of arsenic in drug is not desirable. The Indian pharmacopoeia prescribes the limit for the presence of arsenic as an impurity in various drugs. e.g. sodium chloride should not contain arsenic more than one part per million and sodium citrate should not contain arsenic more than two parts per million.

I. The sample is dissolved in acid which converts the arsenic impurity into arsenious acid or arsenic acid depending upon the valency state of arsenic present in the sample.

$$As^{+3} \longrightarrow As(OH)_3 \text{ or } H_3AsO_3$$

Trivalent - arsenic → Arsenious acid (Trivalent)

$$As^{+5} \longrightarrow O=As(OH)_3 \text{ or } H_3AsO_4$$

Pentavalent arsenic → Arsenic acid (Pentavalent)

II. In presence of reducing agent like potassium iodide (KI), stannous chloride ($SnCl_2$); pentavalent arsenic acid is converted into trivalent arsenious acid.

$$O{=}As(OH)_3 \longrightarrow As(OH)_3$$

$$H_3AsO_4 \text{ ---------------> } H_3AsO_3$$

III. Arsenious acid reacts with nascent hydrogen (which is produced by Zn + HCl) and produces arsenious gas or arsenious hydride.

$$\underset{\text{Arsenious acid}}{H_3AsO_3 + 3H_2} \longrightarrow \underset{\text{Arsine gas}}{AsH_3 + 3\,H_2O}$$

IV. Arsine gas produces yellow stain on the dry mercuric chloride paper. Lead acetate cotton plug traps the H_2S gas formed by sulphidal impurities.

$$\underset{\text{Arsine gas}}{2\,AsH_3 + HgCl_2} \longrightarrow \underset{\text{Yellow stain}}{Hg\,(AsH_2)_2 + 2\,HCl}$$

$$(CH_3COO)_2Pb + H_2S \longrightarrow PbS + 2CH_3COOH$$

The intensity of the standard and sample stain is compared. The intensity of stain with the 'Test solution' should not be more than that produced in the "standard solution".

The apparatus consists of a 100 ml conical flask or bottle closed with a rubber or ground glass stopper through which passes a glass tube (about 20 cm x 5 mm). The lower part of the tube is drawn to an internal diameter of 1.0 mm to 1.5 mm from its tip and has a lateral orifice 2 to 3 mm in diameter.

When the tube is fixed in the stopper bottle the lateral orifice should be at least 3 mm above the surface of the reactants present in the bottle.

The upper end of the tube surface should be at right angle to the axis of the tube. A second glass tube has the same internal diametre and is 30 mm long. It is placed in contact with the first and is held in right position by two spiral springs or clips. Insert loosely packed lead acetate cotton plug in the lower tube. A small square or disc of mercuric chloride paper large enough to cover the orifice of the tube is placed between the flat surfaces.

Separate apparatus is used for the 'Test' and 'Standard'.

Procedure : Introduce the test solution prepared as directed in the individual monograph in the conical flask or bottle. Add 5 ml of 1ml potasium iodide and 10 gm. of Zinc (AsT*)- tested for arsenic impurities. The flask is immediately assembled in a water bath so that a uniform evolution of gas is maintained. After 40 minutes any stain produced on the mercuric chloride paper should not be more intense than that obtained by treating in the same manner with 1.0 ml of arsenic standard solution (10 PPM As) diluted to 50 ml with water.

Quantitative Analysis

Titration : It is a method by which strength of any substance in solution can be determined. If quantity or percentage of pure compound is determined, it is called assay.

Type of Titration

- Direct titration
- Indirect titration
- Non aqueous
- Aqueous
  - Gravimetry
  - Acid - base
  - Redox
    - Nitrite titration
    - Permanganometry
    - Ceric titration
    - Iodine
      - Iodometry
      - Iodimetry
  - Complexometric

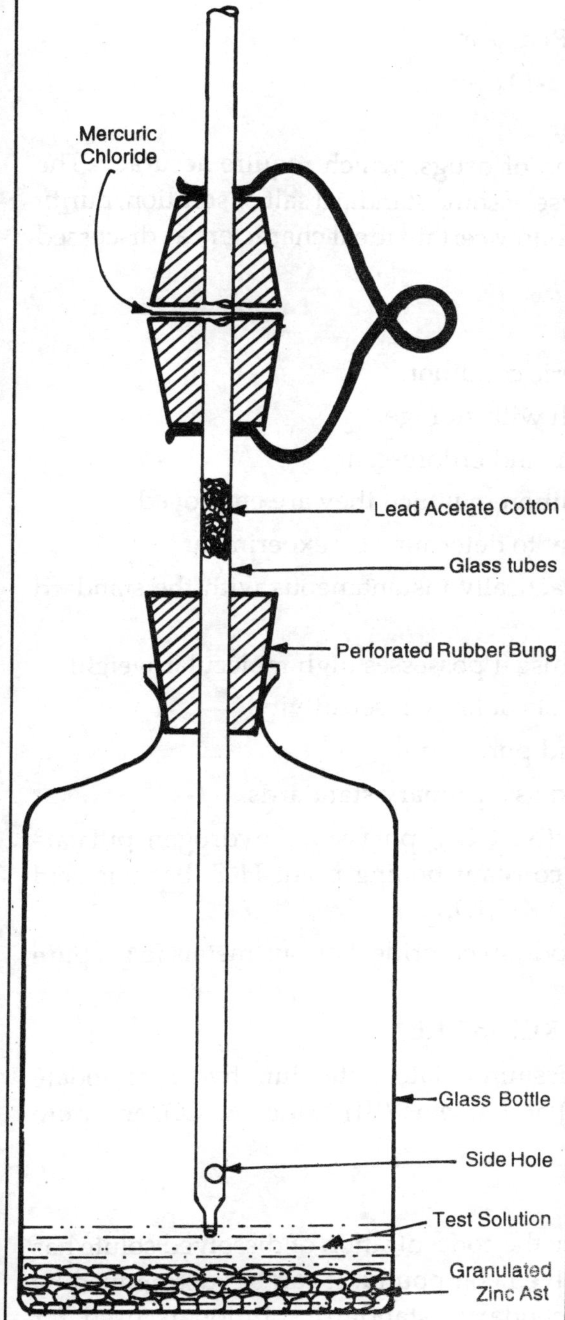

GUTZEIT APPARATUS
(LIMIT TEST FOR ARSENIC)

Direct titration : When titrant used in reaction is directly proportional to amount of substance, it is called direct titration. This method is used when compound is stable and rate of reaction is sufficient.

Indirect titration : When unknown substance is first reacted with some agent, excess of this agent or unreacted agent is determined with the help of third agent,i.e, back titration. These types of titrations are required for volatile substance. The titrations where heating is required for reaction, a parallel titration is conducted omitting the sample. It is called "Blank titration".

Non aqueous titration : This type of titration is performed between very weak acids and bases (according to Lowry Bronsted theory). In non aqueous titration, water acts as acid so it should be prohibited.

Acid base titration : In this titration, acid is titrated with base and point of neutralization, shows the completion of reaction. This is acid base or neutralization titration.

Acid + Base ——> Salt + Water.

Complexometry : This titration is between a metal ion and ligand, in presence of buffer medium. The ligand is selected according to stability of metal ligand complex.

Metal + Ligand ————> Complex

Argenometry : This type of titration involves silver ion and is also called precipitation titration.

$AgNO_3$+Substance——>Precipitation of Ag Salt

Gravimetry titration : The substance is converted from soluble salt into insoluble salt in these titration. Which is filtered, washed, dried and weighed.

Soluble Salt + reagent ————> Insoluble Salt

Redox titration : The titration in which oxidation reduction reaction takes place is called redox titration. These are further divided into various categories as per the involvement of reduced or oxidised ion.

$$Oxi_1 \longrightarrow Red_1 + e^-$$
$$Red_2 + e^- \longrightarrow Oxi_2-$$

Primary Standard

Titrations are performed for quantitative estimation of drugs, which require accuracy. The accuracy should be achieved and confirmed by the use of some standard salt or solution, purity or strength of which should be known and should follow certain ideal characters as discussed below. These agents are called primary standard.

Ideal characteristic of primary standard

1. These agents should be stable with atmospheric condition.
2. These solutions should not alter their strength with storage.
3. They should not be hygroscopic, deliquescent and efflorescent.
4. They should be easily soluble under the condition in which they are employed.
5. The titration error should be negligible or easy to determine by experiment.
6. The reaction should be stochiometric and practically instantaneous with the standard solution.
7. The weighing error should be negligible because it possesses high molecular weight.
8. It should be tested by qualitative and other tests of known sensitivity.
9. It must be easy to obtain, and store in dry, and pure form.

The following substances are commonly employed as a primary standards.

I. Acid base reactions e.g. Sodium carbonate (Na_2CO_3), potassium hydrogen pthalate $KH(C_8H_4O_4)$, sodium tetraborate ($Na_2B_4O_7$), constant boiling point HCl, benzoic acid (C_6H_5COOH) and potassium hydrogen iodide $KH(IO_3)_2$.

II. Complex reactions e.g. Silver nitrate, silver, sodium chloride, various metals (e.g. : pure zinc, Mg, Cu, and Mn) and their salts.

III. Precipitation reaction e.g. NaCl, Ag, $AgNO_3$, KCl and KBr.

IV. Oxidation reduction reactions : $K_2Cr_2O_7$, Potassium iodate, potassium hydrogen iodate [$KH(IO_3)_2$,], $KBrO_3$ sodium oxalate ($Na_2C_2O_4$) and arsenic (III) oxide (As_2O_3), and pure iron.

Secondary standard

Secondary standard solution is a solution in which the concentration of dissolved solute has not been determined from the weight of the dissolved compound. It can be determined by titration against primary standard solution. Secondary standard solution is used for standardisations, and whose content of the active substance has been determined by comparison against a primary standard. e.g. Hydrated salts like $Na_2B_4O_7, 10H_2O$ (Sodium tetra borate) and $CuSO_4, 5H_2O$ (Copper sulphate) are a secondary standards.

2

Common Inorganic Pharmaceutical Compounds

(A) ACIDS AND BASES

Concept of Acids and Bases

There are six theories of acids and bases.

1. Arrhenius concept.
2. Bronsted-Lowry concept.
3. Lewis concept.
4. Dual behaviour of water.
5. Acidity & Basicity of molecules
6. Usanovich concept.

1. Arrhenius concept

The first conceptual definition of acids and bases was given by Arrhenius (1887) in his famous theory of ionisation. According to him an acid is a substance that furnishes hydrogen ions and base is a substance that furnishes hydroxyl ions, when dissolved in water.

$$HCl \longleftrightarrow H^+ + Cl^- \text{ (a strong acid)}$$

$$CH_3COOH \longleftrightarrow CH_3COO^- + H^+ \text{ (a weak acid)}$$

$$NaOH \longleftrightarrow Na^+ + OH^- \text{ (a strong base)}$$

$$NH_4OH \longleftrightarrow NH_4^+ + OH^- \text{ (a weak base)}$$

The properties of acids are due to H^+ ions and bases are due to OH^- ions. An acid which furnishes large concentration of H^+ ions, or high degree of ionisation, is called strong acid. On the other hand, if the degree of ionisation is less, it is called weak acid.

Advantages

(i) When acid and base react with each other and form salt and water, it is commonly known as neutralisation reaction.

$HCl + NaOH \longrightarrow H_2O + NaCl$

Acid Base Water Salt

or $H^+ + OH^- \longrightarrow H_2O$

(ii) This concept has offered a means of co-relating catalytic behaviour with the concentraion of the H^+ion.

Disadvantages

(i) According to this concept, hydrochloric acid is regarded as an acid only when dissolved in water and not in nonpolar solvents such as benzene or when it exists in the gaseous form.

(ii) The neutralization process is limited to those reactions which can occur in aqueous solutions only, although reactions involving salt formation take place in many other solvents and even in the absence of solvents but it is not considered as neutralization reaction.

(iii) Arrhenius concept could not explain the basic nature of substances like ammonia or sodium carbonate which do not contain OH^- ions and acidic nature of substances like CO_2 and SO_2 which do not contain H^+ions.

2. Bronsted-Lowry concept

In 1923 the Danish Chemist, J.N. Bronsted, and the British Chemist, J.M. Lowry, independently defined acids and bases as follows.

"Acid" is a substance that can donate a proton. It is also called proton donor. "Base" is a substance that can accept a proton. It is also called proton acceptor.

Acid - Proton —> Base

or Acid -> Conjugate base + Proton.

Thus acid-base reaction may be written as below

Conjugate Acid Base Pair

$HA + B \longrightarrow HB^+ + A^-$

Acid 1 Base 2 Acid 2 Base 1

Conjugate acid base pair

Some specific examples are given below to explain the theory of Bronsted Lowry concept.

(i) $HCl + H_2O \longrightarrow H_3O^+ + Cl^-$

Acid 1 Base 2 Acid 2 Base 1

(ii) $H_2SO_4 + H_2O \longrightarrow H_3O^+ + HSO_4^-$

(iii) $HSO_4^- + H_2O \longrightarrow H_3O^+ + SO_4^{-2}$

(iv) $HCl(g) + NH_3(g) \longrightarrow NH_4^+ + Cl^-$

(v) $CH_3COOH + H_2O \longrightarrow H_3O^+ + CH_3COO^-$

(vi) $SO_2 + 2H_2O \longrightarrow HSO_3^- + H_3O^+$

(viii) $H_2O + CH_3NH_2 \longrightarrow CH_3^+ NH_3 + OH^-$.

A strong acid has a large tendency to donate a proton. The conjugate base of a strong acid would be a weak base. Thus Cl^-, NO_3^-, SO_4^{2-} etc. are weak bases. The conjugate base of weak acid would be a strong base.

3. Lewis concept

G.N. Lewis further gave a more general definition of acids and bases. According to Lewis :

"Base" is an electron pair donar and "Acid" is an electron pair acceptor. Lewis base is also called nucleophilic, and Lewis acid is called electrophilic substance. Examples of Lewis-bases are Ammonia, Amines, Ethers, Alochols, Water etc. and Examples of Lewis-acids are Boron Trifluoride (BF_3), Aluminium Chloride ($AlCl_3$), Zinc chloride ($ZnCl_2$), Ferric Chloride ($FeCl_3$), Sulphur Trioxide (SO_3), Trimethyl Boron [$(CH_3)_3$ B] etc.

The reaction of a Lewis acid with a Lewis base produces an addition compound. Thus NH_3 with BF_3 is as follows.

(i)

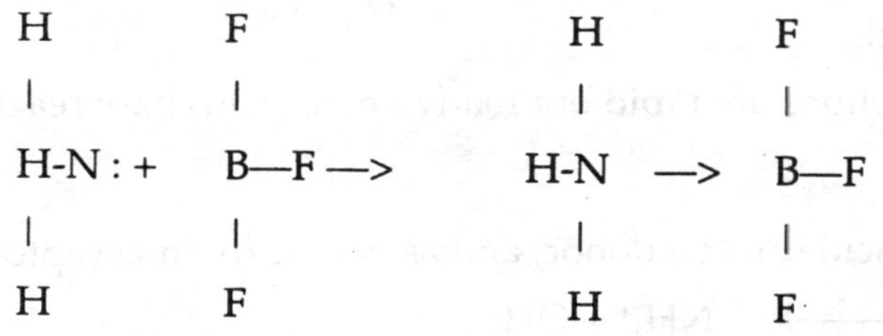

Lewis base Lewis acid addition-compound

(ii)

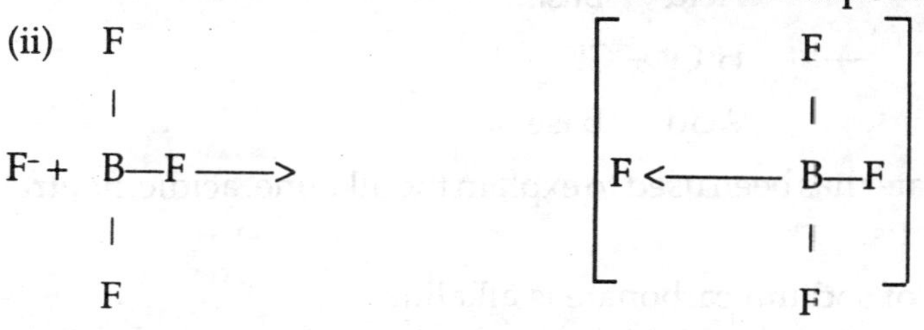

Lewis base Lewis acid Addition Compound

Thus acid base reaction in the Lewis concept is the formation of co-ordinate covalent bond. The process is called co-ordination or neutralisation. Metal ions also act as Lewis acids. They accept electron pair from donor molecules like H_2O, NH_3 etc. or anions like F^-, S^{2-}, CN^- etc.

(i) $2\,H_3N \;+\; Ag^+ \longrightarrow [Ag(NH_3)_2]^+$

Lewis base Lewis acid Addition Compound

(ii) $4\,H_2O \;+\; Cu^{+2} \longrightarrow [Cu(OH_2)_4]^-$

(iii) $2\,CN^- \;+\; Ag^+ \longrightarrow [Ag(CN)_2]^-$

(iv) $6\,F^- \;+\; Al^{+3} \longrightarrow [AlF_6]^{-3}$

Advantages of Lewis concept

(i) This concept also includes those reactions in which no protons are involved.

(ii) Lewis concept is more general than the Bronsted-Lowry concept in which acid-base behaviour is not dependent on the presence of one particular element or on the presence or absence of a solvent.

(iii) It explains the long accepted basic properties of metallic oxides and acidic properties of non- metallic oxides.

(iv) This theory also includes many parameters such as gas phase, high temperature and non-solvent reaction in neutralisation processes.

(v) The Lewis-approach is, however, of great value in cases where the protonic concept is not applicable. e.g. In reactions between acidic and basic oxides in the fused state.

Limitations

1. The acidity and basicity of compound depends on type of reaction hence order of relative strength can not be fixed.
2. Generally acid-base reactions are rapid but many Lewis-acid/base reactions are slow.

4. Dual-Behaviour of water

Water can act both as an acid (Proton donor) and as base (Proton-acceptor).

$H_2O + NH_3 \longrightarrow NH_4^+ + OH^-$

Acid Base Acid Base

$H_2O + HCl \longrightarrow H_3O^+ + Cl^-$

Acid Base Acid Base

The dual-behaviour of water has been used to explain the alkaline, acidic, neutral behaviour of the aqueous solution of salts.

1. Aqueous soln. of sodium carbonate is alkaline.
2. Aqueous soln. of $FeCl_3$, $Al_2(SO_4)_3$, etc. is acidic,

3. Aqueous soln. of NaCl, K_2SO_4 etc. is neutral like:

$H_2O + CO_3^{-2}$	——>	$HCO_3^- + OH^-$
Acid Base		Acid Base (alkaline)
$H_2O + SO_4^{-2}$	——>	$HSO_4^- + OH^-$
Base Acid		Acid Base (acidic)

5. Acidity and Basicity of Molecules :

There are three classes of the acidity and basicity of molecules.

1. Relative acidity of CH_4, NH_3, H_2O and HF is in following order :

   CH_3^- (Strongest base) > NH_2^- > OH^- > F^- (Weakest base)

   CH_4 (Weakest acid) < NH_3 < H_2O < HF (strongest acid)

2. Relative acidity of HF, HCl, HBr, and HI acids is in following order :

   F^- (strongest base) > Cl^- > Br^- > I^- (Weak base)

   HF (Weakest acid) < HCl < HBr < HI (Strongest base)

3. Relative acidity of any acid is in the following order :

   (a) $HCl^{+}O < HCl^{3+}O_2 < HCl_3^{+}O_3 < HCl^{7+}O_4$.

   (b) $H_2S^{4+}O_3 < H_2S^{6+}O_4$.

   (c) $HN^{+}O < HN^{3+}O_2 < HN^{+}O_3$.

   ClO^- (Strongest base) > ClO^{2-} > ClO_3^- > ClO_4^- (Weakest base)

   HClO (Strongest base) < $HClO_2$< $HClO_3^-$ < $HClO_4^-$ (Weakest base)

6. Usanovich Concept

"According to Usanovich concept" an acid is a chemical species which is capable of combining with anions or electrons or giving up cations. Conversely a base is a chemical species which is capable of giving up anions or electrons or combining with cations. e.g.

(i)	$SO_3 + Na_2O$	——>	$Na_2\ SO_4$
	Acid Base		Salt
(ii)	$Fe(CN)_2$ + 4KCN	——>	$K_4[Fe(CN)_6]$
	Acid Base		Salt
(iii)	2 Na	——>	$2\ Na^+ + 2e^-$
	2 Cl + 2 e^-	——>	2 Cl^-
	2 Na^+ + 2Cl^-	——>	2 NaCl .

Advantages

This concept is advantageous in classifying together all examples of acids and base. It is thus the most general theory of all acid-base concepts.

Limitaions

(i) This concept includes all the previous acid-base theories and hence this concept becomes extremely general.

(ii) It includes many reactions such as redox reactions which can be better considered from some other point of view. This theory virtually considers all the chemical reactions as acid-base reactions.

Hydrogen Ion Concentration (pH)

The hydrogen ion concentration, $[H^+]$ in pure water has a value of 10^{-7} moles/litre. The $[OH^-]$ also has the same value. If an acid is added to water, the hydrogen ion concentration is greater than hydroxide ion concentration. Similarly if a base is added to water, the hydroxide ion concentration is greater than hydrogen ion concentration but still they are related by the equation :

$$[H^+][OH^-] = 10^{-14}$$

Thus in neutral solutions $[H^+] = [OH^-]$

in acidic solutions $[H^+] > [OH^-]$

in alkaline solutions $[H^+] < [OH^+]$

A Danish-Chemist "Sorenson" introduced a compact way of representing $[H^+]$ of solution. This is called pH.

The term pH is defined as "negative logarithm of the hydrogen ion concentration".

$$pH = \log [H^+] \text{ or } -\log [H_3O^+]$$

In pure water

$$[H^+] = 10^{-7}$$

$$pH = -\log [10^{-7}] = -\log 1/10^7$$

$$= -\log 1 + \log 10^7$$

$$pH = 7$$

In acidic solution $[H^+]$ is greater than 10^{-7} and hence pH is less than 7.

In alkaline solution $[H^+]$ is smaller than 10^{-7} and pH is greater than 7.

Since $[H^+][OH^-] = 10^{-14}$

or $\log [H^+] + \log [OH^-] = \log 10^{-14} = -14$

or $-\log [H^+] - \log [OH^-] = +14$

or $[pH] + [pOH] = 14$

In pure water $[OH^-] = 10^{-7}$

or $pOH = +7$

It may be noted that since pH is equel to $-\log [H^+]$ or $\log 1/[H^+]$, the pH decreases as the $[H^+]$ increases and vice-versa.

(A-2) Boric Acid

Chemical formula : (H_3BO_3)

Molecular Weight : 61.83

Synonym : Boracic acid.

Standards : It contains not less than 99.5% and not more than 100.5% of boric acid with reference to dried substance.

Preparations

1. *Industrial Process :* It is prepared from the ore colemanite $(Ca_2B_6O_{11}.5H_2O)$. The ore is finely powdered; dissolved in boiling water and sulphur dioxide gas is passed in the solution. The calcium bisulphite is soluble in water and remains in the solution while boric acid crystallizes out.

$$Ca_2B_6O_{11} + 4\,SO_2 + 11\,H_2O \longrightarrow \underset{\text{(Calcium-bisulphite)}}{2\,Ca(HSO_3)_2} + \underset{\text{(Boric-acid)}}{6\,H_3BO_3}$$

2. *Laboratory Process :* Boric acid is prepared by boiling solution of borax (37%) with 20% sulphuric acid. The solution is filtered and crystallized. The crystals of boric acid are separated and dried at room temperature.

$$Na_2B_4O_7 + H_2SO_4 + 5\,H_2O \longrightarrow Na_2SO_4 + 4H_3BO_3$$

Physical Properties

It is a white crystalline colourless shiny plate, uncutaneous to the touch or white crystals, odourless; soluble in boiling water and in 95% ethanol; freely soluble in boiling water, boiling 95% alcohol and glycerin.

Chemical Properties

1. (a) Heating : On heating it losses water and is converted into meta boric acid.

$$H_3BO_3 \xrightarrow{100^0C} HBO_2 + H_2O$$

(b) When it is heated at 160°C it forms pyroboric acid and finally borone trioxide, which has glassy appearance.

$$4\,HBO_2 \xrightarrow{106^0C} \underset{\text{(Pyro boric acid or tetra boric acid)}}{H_2B_4O_7} + H_2O$$

$$H_2B_4O_7 \xrightarrow{\text{more than } 106^0C} \underset{\text{(Borone trioxide)}}{2\,B_2O_3} + H_2O$$

2. With alkali it forms salt

$$NaOH + H_3BO_3 \longrightarrow NaBO_2 + 2\,H_2O$$

(Sod. metaborate)

$$2\,NaOH + 4\,H_3BO_3 \longrightarrow Na_2B_4O_7 + 7\,H_2O$$

Borax

3. Action with calcium flouride and conc. sulphuric acid, it forms boron trifluoride.

$$2\,CaF_2 + 3\,H_2SO_4 + 2\,H_3BO_3 \longrightarrow 3\,CaSO_4 + 2BF_3 + 6\,H_2O$$

4. Action with soda ash liberate carbon dioxide and borax is formed.

$$4\,H_2BO_3 + Na_2CO_3 \longrightarrow Na_2B_4O_7 + CO_2 + 6\,H_2O$$

Borax

5. It reacts with alkali acid fluorides to form sodium fluroborate.

$$2\,NaHF_2 + H_3BO_3 \longrightarrow NaBF_4 + NaOH + 2\,H_2O$$

(Sod. fluroborate).

6. With alcohol : Ethanolic solution of boric acid in ethanol is heated to give ethyl borate. This ethyl borate burns with characteristic green edged flame.

$$H_3BO_3 + 3\,C_2H_5OH \longrightarrow B\,(C_2H_5O)_3 + 3\,H_2O$$

ethyl borate.

Storage : Stored in well closed containers

Assay : It is a very weak acid having pKa (-9.19) for the ionisation of the first proton at 25°C. This lower pKa will not allow to carry out acid base titration. To increase the pKa value the boric acid is esterified with polyhydric compounds such as glycerol, mannitol and catechol. This produces hydronium ion (H_3O^+), which helps in increase of PKa value and acid-base titration can be performed with esterified product.

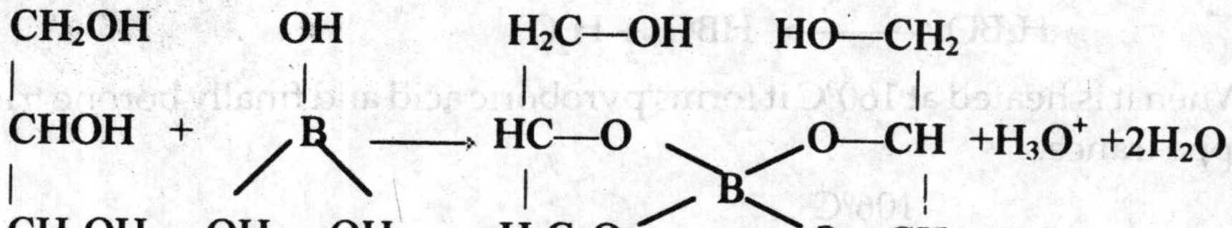

glycero borate

It is assayed by direct acid base type of titration. Weigh accurately about 2gm boric acid and dissolve in 150 ml of 66.6% v/v glycerin solution in water, previously neutralised with phenolphthalein. Titrate the above solution with 1 N NaOH using phenolphtalein as indicator the std. NaOH is standardized with primary std. anhydrous succinic acid.

Factor : Each ml of 1 N NaOH = O. 06183 gm of boric acid.

Uses : The aqueous solutions have been used as mouth washes, gargles, eye lotion, skin lotion, douches for irrigating the bladder and vagina, hot fomentations for ulcers, whitlows, boils, carbuncles and buffer solution for external use.It is dangerous if ingested, hence should not be used for internal preparations.

(A-3) Hydrochloric acid

Chemical formula : HCl

Molecular Weight : 36.46

Standards : It contains not less than 35% w/w and not more than 38% w/w of HCl.

Preparation : A calculated quantity of common salt and conc. sulphuric acid are heated in cast iron pan in salt cake furnace. The hydrochloric acid gas goes down with water spray.

$$NaCl + H_2SO_4 \longrightarrow NaHSO_4 + HCl$$

$$NaHSO_4 + NaCl \longrightarrow Na_2SO_4 + HCl$$

Physical Properties

It is clear, colourless, fuming liquid with pungent odour and miscible with water. Its specific gravity at 20°C is 1.18 gm.

Chemical properties

1. It is dissociated into hydrogen and chloride at 1500°C

$$2\,HCl \longrightarrow H_2 + Cl_2$$

2. It forms dense white fumes with ammonia.

$$NH_3 + HCl \longrightarrow NH_4Cl$$

3. It neutralises oxides and hydroxides while carbonate and bicarbonate, decomposed with evolution of carbon dioxide.

$$MgO + 2\,HCl \longrightarrow MgCl_2 + H_2O$$

$$Na_2CO_3 + 2\,HCl \longrightarrow 2\,NaCl + CO_2 + H_2O$$

$$HCl + NaHCO_3 \longrightarrow NaCl + H_2O + CO_2$$

4. When it is mixed with nitric acid in the ratio of 3:1, it forms aquaregia.

$$3\,HCl + HNO_3 \longrightarrow NOCl + 2\,Cl + 2\,H_2O$$

Purity test

1. *For bromide and iodide :* Dilute 5 ml of HCl with 10 ml of water and add 1 ml of chloroform drop by drop with constant shaking. Chloroform layer should not become brown or violet after addition of chlorinated lime solution.

2. *For free chlorine :* To 60 ml of sample add 50 ml of carbon dioxide free water, 1 ml of 10% potassium iodide solution and 0.5 ml of starch solution. Blue colour produced should disappear by addition of 0.2 ml of 0.01 M Sod.thiosulphate.

3. *For sulphite* : Dilute 1 ml with 10 ml of water and add 0.25 ml of barium chloride solution and 0.5 ml of 0.01 M iodine, the colour of iodine should not completely disappear.

Storage : Stored in glass stoppered containers below 30°C.

Assay : It is a direct acid-base titration. Weigh about 2 gm, add 30 ml of water and titrate with std. 0.1 N sodium hydroxide using methyl red as indicator. The standard sodium hydroxide is standardized with primary standard anhydrous succinic acid.

$$NaOH + HCl \longrightarrow NaCl + H_2O$$

Factor : Each ml of 1 N NaoH = O. 03646 gm of HCl

Uses : Hydrochloric acid is used as an acidifying agent and a pharmaceutical aid.

(A-4) Strong Ammonia

Chemical formula - NH_3

Molocular Weight - 17.03

Standard : It contains not less than 24.5% w/w and not more than 25.5% w/w of ammonia.

Preparation

Haber's Process

When a mixture of nitrogen and hydrogen in the ratio of 1:3 is compressed to a pressure of 200-500 atmosphere and passed over finely divided iron and molybdenum at a temp. about 550°C, Ammonia is formed.

$$N_2 + 3H_2 \xrightarrow[\text{200-500 at. Pr. } 550^0\text{C temp.}]{\text{Cat. Iron+Molybdenum}} 2NH_3$$

Physical Properties : It is clear colourless liquid, strong pungent, with characteristic odour; and miscible with water.

Chemical Properties

1. It reacts with acid to form salt.

$$NH_3 + HCl \longrightarrow NH_4Cl.$$

2. It readily forms soluble complex with cations of Ag, Cu, Au, Zn, Cd, Cr, Ni, Co, Mn and Pt.

$$AgCl + 2NH_3 \longrightarrow [Ag(NH_3)_2]^+Cl^-$$

3. It forms basic ammonium salt which is similar to product obtained by Hydrolysis of salts.

$$HgCl_2 + H_2O \longrightarrow Hg(OH)Cl + HCl$$
$$HgCl_2 + NH_3 \longrightarrow Hg(NH_2)Cl + HCl$$

Incompatibility : Ammonia is incompatible with iodine, hypochlorites, salts of heavy metals (e.g. mercuric chloride and silver salts.), alkaloidal salts and tannins.

Storage : It should be stored in airtight containers, in a cool place.

Assay : It is assayed by indirect (back) type of acid-base titration. Weigh accurately about 3 gm of ammonia in a flask containing 50 ml of std. 1 N sulphuric acid. The excess of acid is back titrated with standard 1 N sodium hydroxide, using methyl red as indicator. The 1 N sodium hydroxide is standardised with anhydrous primary standard succinic acid.

$$2\,NH_4OH + H_2SO_4 \longrightarrow (NH_4)_2SO_4 + 2H_2O$$

$$H_2SO_4 + 2\,NaOH \longrightarrow Na_2SO_4 + 2H_2O$$

Factor : Each ml of 1 N H_2SO_4 = 0.01703 gm of ammonia

Uses : When ammonia is inhaled, it irritates mucosa of upper respiratory tract and stimulates respiration, which accelerates heart and increses blood pressure. It is applied externally for counter irritant and rubefacient action. Its salts (NH_4Cl) are used as expectorant and diuretic.

(A-5) Calcium hydroxide

Chemical formula : $Ca(OH)_2$

Molecular Weight : 74.09.

Synonym : Slaked lime, Hydrated lime.

Preparation

Calcuim hydroxide is prepared by absorbing water over calcium oxide. It is a exothermic reaction and calcium oxide lumps are made to swell and disintegrated to fine powder. The quality of calcium hydroxide depends on the quality of raw material.

$$CaCO_3 \xrightarrow{1075^\circ K} CaO + CO_2$$

$$CaO + H_2O \longrightarrow \underset{\text{(calcium hydroxide)}}{Ca(OH)_2}$$

Physical Properties

It is a soft white powder, with a slightly bitter taste; soluble in 600 parts of water, less soluble in hot water, more soluble in aqueous solution of sugar and glycerol.

Chemical Properties

1. On heating looses water and forms calcium oxide.

$$Ca(OH)_2 \longrightarrow \underset{\text{(calcium oxide)}}{CaO} + H_2O$$

2. On reaction with inorganic acid; it form salts.

$$Ca(OH)_2 + 2\,HCl \longrightarrow CaCl_2 + 2H_2O$$

3. When carbon dioxide is passed in water, it forms carbonic acid. This carbonic acid reacts with lime water $[Ca(OH)_2]$ to form calcium carbonate.

$$H_2O + CO_2 \longrightarrow H_2CO_3$$

$$Ca(OH)_2 + H_2CO_3 \longrightarrow \underset{\text{calcium carbonate}}{CaCO_3} + 2\,H_2O$$

Storage : Stored in air-tight well filled containers.

Uses : It is used as astringent and antacid. It is also used in the formation of calcium soaps with fatty acids, which act as emulsifying agent. Its solution is used to absorb carbon dioxide gas, thus act as pharmaceutical aid.

(A-6) Sodium hydroxide

Chemical formula : NaOH

Molecular Weight : 40.00

Synonym : Caustic soda

Standards : It contains not less than 97.0% and not more than 100.5% of NaOH.

Preparations

1. Electrolytic-process (Castner-Kellner process)

A saturated solution of sodium chloride (brine) is electrolysed employing carbon anode and metallic cathode. The cathode compartment should be brine free so that sodium hydroxide can be collected. At anode, valuable gases like Hydrogen and chlorine are collected.

$$NaCl \longleftrightarrow Na^+ + Cl^-$$

$$H_2O \longleftrightarrow OH^- + H^+$$

NaOH (At cathode) HCl (At anode)

2. From sode lime

The sodium carbonate is heated with milk of lime in large tanks. It produces sodium hydroxide and calcium carbonate. This reaction is reversible hence excess quantity of sodium carbonate is used.

$$Na_2CO_3 + Ca(OH)_2 \longleftrightarrow CaCO_3 + 2\,NaOH$$

Physical properties

It is a white crystalline masses, available in flakes or pellets, deliquescent; absorbs atmospheric carbon dioxide; very soluble in water and soluble in 95% ethanol.

Chemical Properties

1. Action with non-metals

(a) With halogens it forms hypohalites in cold condition and forms halates in hot condition.

$$Cl_2 + 2NaOH \text{ (cold)} \longrightarrow NaCl + NaClO + H_2O$$

Sod. hypo. chlorate

$$3Cl_2 + 6NaOH \text{ (hot)} \longrightarrow 5NaCl + NaClO_3 + 3H_2O$$

Sod. chlorate

(b) With sulphur it gives sodium Sulphide and sodium thiosulphate

$$6\ NaOH + 4\ S \longrightarrow 2\ Na_2S + Na_2S_2O_3 + H_2O$$

Sod. sulphide Sod. thio sulphate

(c) With phosphorous it gives phosphine gas.

$$4\ P + 3\ NaOH + 3\ H_2O \longrightarrow 3\ NaH_2PO_2 + PH_3$$

(Phosphine)

(d) With silicon it form sodium silicate and hydrogen gas.

$$Si + 2\ NaOH + H_2O \longrightarrow Na_2SiO_3 + 2\ H_2$$

2. Action with metals

It reacts with metals like zinc,aluminum and tin to liberate hydrogen gas.

$$Zn + 2\ NaOH \longrightarrow Na_2ZnO_2 + H_2$$

Sod. zincate

$$2\ Al + 2\ NaOH + 2\ H_2O \longrightarrow 2\ NaAlO_2 + 3H_2$$

Sod. meta aluminate

$$Sn + 2\ NaOH + H_2O \longrightarrow Na_2SnO_3 + 2H_2$$

Sod. stannate

3. Action with acids

It reacts with acids to form salts.

$$NaOH + HCl \longrightarrow NaCl + H_2O$$

$$2\ NaOH + H_2SO_4 \longrightarrow Na_2SO_4 + 2H_2O$$

$$NaOH + HNO_3 \longrightarrow NaNO_3 + H_2O$$

$$3\ NaOH + H_3PO_4 \longrightarrow Na_3PO_4 + 3H_2O$$

$$NaOH + CH_3COOH \longrightarrow CH_3COONa + H_2O$$

4. Action with ammonium salts : When heated with ammonium salt and it liberate ammonia gas.

$$NH_4Cl + NaOH \longrightarrow NaCl + H_2O + NH_3$$

$$2\,NaOH + (NH_4)_2SO_4 \longrightarrow Na_2SO_4 + 2\,NH_3 + 2\,H_2O$$

5. Action on metallic salt : It reacts with various metal salts of Mn, Ni, Cu, to form stable hydroxides while with Ag, Zn, Al, it forms unstable hydroxides.

(I)

$$MnSO_4 + 2\,NaOH \longrightarrow Mn(OH)_2 + Na_2SO_4$$

$$Ni(NO_3)_2 + 2\,NaOH \longrightarrow Ni(OH)_2 + 2\,NaNO_3$$

$$FeSO_4 + 2\,NaOH \longrightarrow Fe(OH)_2 + Na_2SO_4$$

$$CuSO_4 + 2\,NaOH \longrightarrow Cu(OH)_2 + Na_2SO_4$$

(II)

$$AgNO_3 + NaOH \longrightarrow Ag(OH) + NaNO_3$$

$$2\,AgOH \longrightarrow Ag_2O + H_2O$$

$$ZnCl_2 + 2\,NaOH \longrightarrow Zn(OH)_2 + 2\,NaCl$$

$$Zn(OH)_2 + NaOH \longrightarrow Na_2\,Zn\,O_2 + 2\,H_2O$$

$$AlCl_3 + 3\,NaOH \longrightarrow Al(OH)_3 + 3\,NaCl$$

$$Al(OH)_3 + NaOH \longrightarrow NaAlO_2 + 2\,H_2O$$

Storage : It should be stored in air tight containers; if it is stored in glass containers then it should be closed by wax, corks or plastic-lined screw cap.

Assay : It is assayed by direct acid base titration. Weigh accurately about 2 gm of substance and dissolved in 80 ml of freshly boiled and cooled (free from carbon dioxide) water, titrate with 1 N HCl, using methyl orange as indicator.

Factor : Each ml of 1 N HCl = 0.04 gm of sodium hydroxide.

Uses : It is used as cuticle solvent and as pharmaceutical aid.

(A-7) Buffer Solutions

A solution resisting the change in pH is known as a buffer solution. When solution resists the change in pH when acid or alkali is added, such action is called as a buffer action. A buffer solution has a reserved pH or a fairly constant pH, even when small amount of acid or alkali is added to it.

Types of Buffer solutions : Buffer solution are of two types.

(1) Acidic buffer solution : A weak acid and its salt with strong base. e.g. acetic acid and sodium acetate ($CH_3COOH + CH_3COO^- Na^+$)

(2) Basic buffer solution : A weak base and its salt with a strong acid. e.g. ammonium hydroxide and ammonium chloride ($NH_4OH + NH_4Cl$).

pH of acidic buffer solution

The hydrogen ion concentration or pH obtained from the dissociation of weak acid (HA), is given by the equation.

$$HA \longleftrightarrow H^+ + A^-$$

$$\frac{[H^+][A^-]}{[HA]} = Ka$$

$$\text{or} \quad [H^+] = K_a \frac{[HA]}{[A^-]}$$

$$\text{or} \quad -\log[H^+] = -\log K_a - \log \frac{[HA]}{[A^-]}$$

$$\text{or} \quad pH = pK_a + \log \frac{[A^-]}{[HA]}$$

$$pH = pK_a + \log \frac{[\text{conjugate base}]}{[\text{acid}]}$$

This equation is called as "Henderson-Hasselbach equation".

pH of alkaline buffer solution

The hydrogen ion concentration (pH) obtained from the dissociation of weak base, is given by the equation :

$$BOH \longleftrightarrow B^+ + OH^-$$

$$K_b = \frac{[B^+][OH^-]}{[BOH]}$$

$$\text{or} \quad [OH^-] = K_b + \log \frac{[BOH]}{[B^+]}$$

$$\log[OH^-] = \log K_b + \log \frac{[BOH]}{[B^+]}$$

$$\text{or} \quad -\log[OH^-] = -\log K_b + \log \frac{[B^+]}{[BOH]}$$

$$\text{or} \quad p[OH] = pK_b + \log \frac{[\text{conjugate acid}]}{[\text{Base}]}$$

$$pH = 14 - p[OH]$$

$$pH = 14 - pK_b + \log \frac{[\text{conjugate acid}]}{[\text{Base}]}$$

$$\text{or} \quad pH = 14 - pK_b + \log \frac{[\text{salt}]}{[\text{base}]}$$

This equation is called as "Henderson Hasselbatch equation"

Buffer Capacity

Buffer solution resists change in pH upon addition of strong acid or strong base. The buffering action is calculated in terms of buffer-capacity. Buffer-capacity is defined as the moles of strong acid or strong base required to change the pH of one litre of buffer solution by one unit. The larger the buffer capacity, the better is the buffer, since it can accomodate more acid or base without significant change in pH. As a solution with 0.01 N of acetic acid and sodium acetate has the same pH as that of solution with 0.1 N of these constituents, the later solution can consume ten times as much strong acid or strong base for the same change in pH. Thus higher, the buffer capacity the more effective in controlling change in pH.

[B] ANTI OXIDANTS

Antioxidants are the agents, capable of functioning as reducing agent and themselves are easily oxidised in pharmaceutical preparation. They act either by getting oxidised in place of active pharmaceutical ingredients or if active ingredient is oxidised, they reduce the oxidised active ingredient. Oxidation and reduction are simultaneous process. In this process, loss or gain of electron are always balanced in a complete reaction.

$$OX_1 + \bar{e} \longleftrightarrow Red_1$$

$$Red_2 \longleftrightarrow OX_2 + e^-$$

$$OX_1 + Red_2 \longleftrightarrow Red_1 + OX_2$$

The efficiency of chemical substance to undergo redox-reaction can be determined by measuring electrode potential with the help of "Nernst's equation".

$$E_{cell} = E^o_{cell} - \frac{RT}{nF} \ln Q$$

$$\text{or} \quad E_{cell} = E^o_{cell} - \frac{0.059}{n} \log Q$$

Where (RT/F=0.059)

$$\text{or} \quad E_{cell} = E_{1/2}\,\text{oxid} + E_{1/2}\,\text{red.}$$

Selection of anti-oxidants : The selection of anti-oxidants for pharmaceutical purposes is based on following parameters.

1. Desired redox reaction should be assessed by the use of standard electrode potential and by the "Nernst-equation".
2. Substance should be chemically compatible with other ingredients.
3. It should not have any physiological action.
4. It should be active in low-concentration.
5. It should not produce any toxicity even after oxidation or reduction.

These anti-oxidants may cause serious problem when combined together because some strong agents may form explosive mixtures when combined in dry form or in concentrated solutions.

(B-1) Hypo-Phosphorus acid

Chemical formula : H_3PO_2

Molecular Weight : 65.99

Preparations

1. It is prepared by the treatment of potassium hypophosphite with tartaric acid.

$$KH_2PO_2 + C_4H_6O_6 \longrightarrow H_3PO_2 + KHC_6H_4O_6$$

Pot.hypo. phosphite — Tartaric acid — Hypophos -phorus acid — Pot.bitartrate

2. When calcium hypophosphite is treated with oxalic acid or sulphuric acid. It forms hypophosphorus acid and insoluble calcium oxalate or calcium sulphate.

$$Ca(H_2PO_4)_2 + H_2C_2O_4 \longrightarrow CaC_2O_4 + 2H_3PO_2$$

Cal. hypo — oxalic acid — Cal. oxalate

$$Ca(H_2PO_4)_2 + H_2SO_4 \longrightarrow CaSO_4 + 2H_3PO_2$$

Cal. phosphite

Physical Properties

It is a clear or yellowish liquid, with slight odour, strong acidic taste, soluble in water and alcohol having specific gravity 1.32 at 25°C.

Chemical Properties

It reduces iodine and acidic $KMnO_4$ solution immediately.

$$H_3PO_2 + 2I_2 + 2H_2O \longrightarrow 4HI + H_3PO_4$$

$$5H_3PO_2 + 4KMnO_4 + 6H_2SO_4 \longrightarrow 2K_2SO_4 + 4MnSO_4 + 5H_3PO_4 + 6H_2O$$

Incompatibility : When it is kept with nitrates, chlorates and permagnates, it forms explosives.

Storage : It should be stored in well-closed containers in a dark place.

Uses : It acts as monoprotic (only one ionisable-hydrogen) acid and is added in dilute hydroiodic acid, ferrous iodide syrup and hydroiodic acid syrup as powerful anti-oxidant. It passes out from the body without any change and is regarded as brain tonic.

(B-2) Sulphur dioxide

Chemical Formula : SO_2

Molecular Weight : 64.06

Preparation

On laboratory scale sulphur dioxide is prepared by burning sulphur in air or in oxygen atmosphere.

$$S + O_2 \longrightarrow SO_2$$

Physical Properties

It is a colourless, non-flammable, pungent, and an irritant-odour and one ml of sulphur dioxide dissolves in 32.8 ml of water. Its boiling point is -10°C.

Chemical Properties

1. It forms addition product with halogens in presence of sun-light or camphor.

$$SO_2 + Cl_2 \longrightarrow SO_2Cl_2$$

Sulphuryl chloride

$$SO_2 + F_2 \longrightarrow SO_2F_2$$

Sulphuryl fluoride

2. It acts as reducing agent and reduces iodine to iodide ions, permagnate to manganese salts (decolorisation) and iodates to iodine etc.

$$I_2 + 2H_2O + SO_2 \longrightarrow H_2SO_4 + 2HI$$

$$2KMnO_4 + 2H_2O + 5SO_2 \longrightarrow K_2SO_4 + 2MnSO_4 + 2H_2SO_4$$

$$2KIO_3 + 4H_2O + 5SO_2 \longrightarrow 2KHSO_4 + I_2 + 3H_2SO_4$$

Incompatibility : In moderate to strong acidic pH, sulphur dioxide is stable. Sulphur dioxide is converted totally to disulphur and sulphite, when the pH changes from neutral to alkaline values, in presence of buffer solutions, the SO_2 reduces OH^- ion concentrations and causes a shift to more acidic pH.

$$SO_2 + OH^- \longrightarrow HSO_3^-$$

Uses : Sulphur dioxide is gaseous in nature. Hence used in single dose injectables as anti-oxidants. Sulphur dioxide in glycerin is used for sour throat, tonsilitis and skin-infection.

(B-3) Sodium bisulphite

Chemical formula : $NaHSO_3$

Molecular Weight : 104.06

Synonyms : Sodium hydrogen sulphite, Sodium acid sulphite, Sodium pyrosulphate.

Preparation

It is prepared by passing sulphur dioxide into a solution of sodium carbonate till solution gets saturated.

$$Na_2CO_3 + H_2O + SO_2 \longrightarrow 2NaHSO_3 + H_2 + CO_2$$

Physical Properties

It is white crystalline powder; pungent irritating odour; freely soluble in water and slightly soluble in alcohol.

Chemical Properties

When it is treated with hydrochloric acid, it forms sulphurous acid (solution of sulphur dioxide in water).

$$NaHSO_3 + HCl \longrightarrow NaCl + \underset{\text{Sulphurous acid}}{H_2SO_3}$$

Uses : It is used as anti-oxidant, stabilizing agent and antiseptic. It's 0.2% solution is anti-oxidant for epinephrine-hydrochloride, phenylephrine hydrochloride and ascorbic acid injections. It is antiseptic in gastric fermentation and externally for parasitic skin diseases. This is also useful as reducing agent for solubilizing dyes, permagnate stains remover etc.

(B-4) Sodium Metabisulphite

Chemical formula : $Na_2S_2O_5$

Molecular Weight : 190.10

Synonyms : Disodium pyrosulphite, Sodium pyrosulphite.

Standards : It contains not less than 95% and not more than 100.5% of $Na_2S_2O_5$.

Preparations

Sulphur dioxide gas is passed over hot concentrated sodium hydroxide solution until it is saturated.

$$NaOH + SO_2 \longrightarrow NaHSO_3$$

$$2NaHSO_3 \longrightarrow \underset{\text{Sod. metabisulphite}}{Na_2S_2O_5} + H_2O$$

The sodium bisulphite salt (unstable) crystallize out which on heating separates sodium metabisulphite (stable).

Physical Properties

It is a colourless, prismatic, white or creamy white crystalline powder; with sulphurous odour; saline taste; freely soluble in water and slightly soluble in 95% alcohol.

Purity test

1. **Iron :** Add 2ml of hydrochloric acid to 0.5 gm of substance (sodium metabisulphite) and evaporate in a water bath to dryness. Dissolve the residue in 2 ml of hydrochloric acid and 20 ml of water and add a few drops of bromine solution, cool, dilute with water to 25 ml than add 50 mg of ammonium persulphate and 5 ml of ammonium thiocynate solution. Any colour produced should not be more intense than that of obtained by adding 5 ml of ammonium thiocynate solution to a mixture of iron standard solution (20 PPM Fe), 2 ml of HCl, 50 mg of ammonium persulphate and sufficient water to produce 25 ml (40 PPM).

2. **Thiosulphate :** To 1.0 gm in 10 ml of 2 M hydrochloric acid and heat on a water bath for 10 minutes; not more than a faint opalescence is produced..

Uses : It is an anti-oxidant and reducing agent. This is widely employed in pharmaceutical preparations, especially in those containing substances which are readily oxidised to form coloured decomposed products. A chelating agent such as sodium-edetate is used in conjunction with sodium meta-bisulphite to remove heavy metallic ions which often catalyses oxidation reaction. It is also used to stabilize injections containing salts of adrenaline or morphine and also is used as preservative in acidic solutions and syrups.

(B-5) Sodium thiosulphate

Chemical Formula : $Na_2S_2O_3.5H_2O$

Molecular Weight : 248.17

Synonyms : Sodium Hyposulphite, Hypo, Natrii-Thiosulfas, Antichlor.

Standards : It contains not less than 99.0% and not more than 101.00% of $Na_2S_2O_3.5H_2O$

Preparations

When sodium sulphite is heated with powdered sulphur it gives sodium thiosulphate.

$$Na_2SO_3 + S \longrightarrow Na_2S_2O_3$$

Sodium thiosulphate

Physical Properties

It is an odourless; colourless; monoclinic prisms; deliquescent in moist air and efflorescence in dry air. It is very soluble in water but practically insoluble in alcohol.

Chemical Properties :

1. It reacts with dilute acid to form sulphur dioxide and sulphur.

$$Na_2S_2O_3 + 2HCl \longrightarrow 2NaCl + H_2S_2O_3$$

$$H_2S_2O_3 \longrightarrow Na_2O + SO_2 + S$$

2. When it is heated at 225°C it gets converted to sodium pentasulphide.

$$4\ Na_2S_2O_3 \xrightarrow{225^0C} 3\ Na_2SO_4 + Na_2S_5Na_2S_5 \longrightarrow Na_2S_4 + S$$

3. It reacts with silver nitrate to form black precipitate of silver sulphide.

$$2AgNO_3 + Na_2S_2O_3 \longrightarrow 2\ NaNO_3 + Ag_2S_2O_3$$

$$\underset{\text{White}}{Ag_2S_2O_3} + H_2O \longrightarrow \underset{\text{Black}}{Ag_2S} + H_2SO_4$$

4. When it is treated with photographic film, it removes unreacted silver bromide by forming a complex. This property is called fixing property of photographic film.

$$AgBr + Na_2S_2O_3 \longrightarrow NaAgS_2O_3 + NaBr$$

5. It reacts with halogens to precipitate sulphur

$$Na_2S_2O_3 + Cl_2 + H_2O \longrightarrow Na_2SO_4 + HCl + S$$

$$2\ Na_2S_2O_3 + I_2 \longrightarrow \underset{\text{Sod. tetrathionate}}{Na_2S_4O_6} + 2\ NaI$$

6. It reacts with neutral $KMnO_4$ to form sodium sulphate.

$$Na_2S_2O_3 + 2\ KMnO_4 \longrightarrow Na_2SO_4 + K_2SO_4 + 2MnO_2$$

7. It reacts with barium chloride solution and calcium chloride solution.

$$Na_2S_2O_3 + BaCl_2 \longrightarrow \underset{\text{(White)}}{BaS_2O_3} + 2NaCl$$

$$Na_2S_2O_3 + CaCl_2 \longrightarrow CaS_2O_3 + 2NaCl$$

8. It gives violet colour with ferric chloride

$$3Na_2S_2O_3 + 2FeCl_3 \longrightarrow Fe_2(S_2O_3)_2 + 6NaCl$$

Storage : It should be stored in air tight containers.

Uses : It is used as reducing agent in medicines, in the concentration of 0.01 to 1%, topically acting as antifungal in skin infection, antichlor in bleaching operation, fixer in photographic work and given in cyanide poisoning with sodium nitrite.

Assay : It is assayed by direct redox iodometric type of titration. Weigh accurately about 0.5 gm and dissolve in 20 ml of water. This solution is titrated with 0.5 M iodine solution using starch as indicator. The indicator is added at the end of titration. The iodine solution is standardized by using anhydrous arsenic trioxide as primary standard.

Factor : Each ml of 0.05 M iodine ≡ 0.02482 gms of $Na_2S_2O_3.5H_2O$

$$2\,Na_2S_2O_3 + I_2 \longrightarrow Na_2S_4O_6 + 2\,NaI$$

Use : It is used as pharmaceutical aid.

(B-6) Nitrogen

Chemical formula : N_2

Molecular Weight : 28.01

Preparation

Atmosphere contains 79% of nitrogen gas. The liquid nitrogen is obtained by fractional distillation of liquified air. It is filled in metallic cylinders, painted grey with black on the neck and shoulders, having name and symbol on it.

It also occurs in proteins, amino acids, alkaloids etc. found in various parts of plants, animals and it is also found deposited at Chile saltpetre.

Physical properties

It is an odourless, tasteless, colourless, non-inflammable; soluble in 65 parts of water and a part of alcohol at 20ºC.

Chemical properties

It is an inert gas but combines with Li, B, Si, Ba, Ca, Sr, Mg, oxygen, hydrogen etc. at very high temperature.

1. At high tension electric arc nitrogen combines with atmospheric oxygen.

$$N_2 + O_2 \longleftrightarrow \underset{\text{nitrous oxide}}{2NO}$$

$$2NO + O_2 \longrightarrow 2NO_2$$

$$4NO_2 + 2H_2O + O_2 \longrightarrow 4HNO_3$$

2. When it is heated with three parts of Hydrogen at 550ºC and 200-500 At. Pr. it forms ammonia.

$$N_2 + 3H_2 \xleftrightarrow[550^{\circ}C]{200\text{-}500\text{ At. Pr.}} 2\,NH_3$$

3. When nitrogen is heated with calcuim carbide at 800-1000ºC it forms nitrolim.

$$CaC_2 + N_2 \longrightarrow \underset{\text{Nitrolim}}{[CaCN_2 + C]}$$

4. It forms metal nitrides with Mg and Al at high temp.

$$3Hg + N_2 \longrightarrow Mg_3N_2$$

$$2Al + N_2 \longrightarrow 2AlN$$

Uses : Nitrogen is an inert (very less reactive) gas, used as the substituent of atmospheric oxygen or air, wherever product is susceptible to oxidation. e.g. preparation containing cod-liver oil, castor oil, olive oil, multiple vitamin, etc. are prone to oxidation, hence their injections are packed in nitrogen atmosphere. It is also used as diluent for oxygen gas before administering to patient.

Nitrogen is also used in manufacturing of ammonia, nitrates, cyanides, urea, fertilizers, food freezing process and coolant in laboratory.

(C) GASTRO-INTESTINAL AGENTS

Various inorganic agents are used in the treatment of gastro-intestinal disorders. These are :

1. AntacidAcidifying agent (Alter gastric pH),
2. Protectives (Protect intestinal mucosa),
3. Adsorbents (Adsorbes intestinal toxins), and
4. Laxative or Cathartic (Relieves constipation).

Many of these indication are symptoms of serious condition.

(C-I) Acidifying agents

The hydrochloric acid secreted in stomach performs important functions by killing the bacteria ingested with food and drink. It provides acidic medium and promotes formation of proteolytic enzyme, pepsin. Defect in secretion of hydrochloric acid may cause several diseases.

Achlorhydria is the absence of hydrochloric acid in the gastric secretion. This secretion of hydrochloric acid may or may not be stimulated by histamine phosphate.

The patients suffering from chronic nephritis (inflammation of the kidneys), chronic alcoholism, tuberculosis, hyperthyrodism, pellagra, parasitic infestations and sprue(periodic fatty, frothy diarrhea) may suffer from achlorhydria and this can improve with the histamine phosphate.

The patients suffering from sub total gastrectomy, atrophic gastritis (Chronic gastritis with atrophy of the mucous membranes and glands), carcinoma of the stomach, or gastric polyps do not stimulate hydrochloric acid secretion even after histamine phosphate. In such conditions acidifying agents are used to treat achlorhydria.

(C-I.1) Dilute Hydrochloric acid

Chemical formula : HCl

Molecular Weight : 36.5

Standards : It contains not less than 9.5% and not more than 10.5% w/w of HCl

Preparation

It is prepared by diluting the concentrated Hydrochloric acid with water.

Hydrochloric acid 274.0 gm

Purified water 726.0 gm

Physical Properties

It is clear, colourless, pungent and freely miscible with water.

Chemical Properties

1. It dissociates into hydrogen and chlorine at 1500°C.

   $2\,HCl \longrightarrow H_2 + Cl_2$

2. It forms dense white ammonium chloride clouds with ammonia.

   $NH_3 + HCl \longrightarrow NH_4Cl$

3. It neutralises oxide and hydroxides, while carbonates and bicarbonates, decomposes with evolution of CO_2.

   $MgO + 2\,HCl \longrightarrow MgCl_2 + H_2O$

   $Na_2CO_3 + 2\,HCl \longrightarrow 2\,NaCl + CO_2 + H_2O$

   $NaHCO_3 + HCl \longrightarrow NaCl + H_2O + CO_2$

4. It forms aquaregia with nitric acid in the ratio of 3:1.

   $3HCl + HNO_3 \longrightarrow NOCl + 2Cl + 2H_2O$

Storage : Stored in stoppered containers below 30°C.

Assay : It is assayed by direct acid-base titration. 6 gm of solution is weighed, diluted with water and titrated with 1.0 N std. sodium hydroxide solution using methyl red as indicator. The titrant 1.0 N sodium hydroxide is standardized against anhydrous primary standard succinic acid.

$NaOH + HCl \longrightarrow NaCl + H_2O$

Factor : Each ml of 1.0 N NaOH ≡ 0.03646 gm of HCl.

Uses : Dilute Hydrochloric acid is used in the treatment of achlorhydria, hypochlorhydria and gastrogenous diarrhoea. It act as bacteriocidal in the stomach and changes the pH of substances present in stomach so that enzyme pepsin can act on it.

Dose : Not more than 20 ml per day.

(C-II) Antacids

Antacids are chemical substances which react with gastric acid and lower the acidity of gastric contents. These drugs relieve pain that occurs in hyperchlorhydria, peptic ulcer, heart burn and also reduce spasm in addition to relief from the uncomfortable feeling from over eating and growing hunger feeling between meals. Antacids are weak bases and they raise the gastric pH above 4 by neutralising excess gastric hydrochloric acid. The hydrochloric acid may be causing pain and ulceration. Antacids also inactivate the proteolytic enzyme, pepsin. Generally antacids are free from side effects but a few drugs cause systemic alkalosis. Patients suffering from peptic ulcers are best managed in addition to antacids with organic anti-ulcer agents which are potential histamine release inhibitors. Depending on the severity and location of an ulcer (gastric, duodenal or Jejunal, oesophageal) treatment will range from proper diet, antacid and or anti-ulcer agent and or anticholinergic-agent, patients may be advised to take meals at frequent intervals and avoid acid stimulants, irritating foods. e.g. spices, fried foods, uncooked roughage.

The neutralizing-capacity of an antacid is expressed in milliequivalents (mEq.) of hydrochloric acid. Every antacid product must have a total neutralising capacity of atleast 5 mEq. of hydrochloric acid per dosage unit. It is enough to raise pH of empty stomach to 3.5. If the product is more potent, then it neutralises more acid. If the product which neutralises 2.5 mEq. acid per dose is half as potent as one that neutralises 5 mEq, the one that neutralises 10 mEq is twice as potent.

Antacids can be classified into two types.

A. Absorbable or systemic antacids

B. Non-absorbable or non-systemic antacids.

A. **Systemic antacids** : They are readily absorbable, soluble, and produce systemic electrolytic-alteration and alkalosis e.g. Sodium bicarbonate

B. **Non-absorbable antacids** : These antacids are not absorbed and do not produce systemic electrolytic alteration. These antacid are divided into four subgroups :

a. Aluminium containing antacids : e.g. Aluminium phosphate, Aluminium hydroxide, Basic aluminium carbonate (gel), Dihydroxy aluminium sodium carbonate, Dihydroxyl aluminium amino acetate

b. Calcium containing antacids : e.g. Calcium carbonate, Tribasic calcium phosphate.

c. Magnesium containing antacids : e.g. Magnesium carbonate, Magnesium citrate, $Mg(OH)_2$, MgO, $Mg_3(PO_4)_2$, Magnesium-trisilicate.

d. Combination antacids preparations : e.g. Aluminium hydroxide gel and Magnesium hydroxide, Aluminium hydroxide gel and Magnesium trisilicate, Magaldrate (mono alum hydrate and hydrated magnesium aluminate), Alginic acid and Sodium bicarbonate containing mixture

Ideal Characteristics of antacids

An antacid should posses a few ideal characteristics. These are as follows :

1. The antacid should not produce laxative action or cause constipation.
2. The antacid should not be absorbed or cause systemic alkalosis.
3. The antacid should have quick and prolong action.
4. The antacid should act as buffer in the pH range of 4-6.
5. The antacid should not liberate carbon dioxide when reacting with gastric hydrochloric acid and should not cause rebound hyper acidity.
6. The antacid should be palatable and inexpensive (Cheap).
7. The antacid should not interfere with absorption of food.

Combined Antacid therapy

Antacids are alkaline bases, which neutralize the excess of hydrochloric acid associated with gastritis and peptic ulcers. The antacids act by inactivation of pepsin, which functions optimally at low pH. Antacid given with the chronic administration of any drug, is not without side-effect. It is quite normal for the stomach to be at pH 1-2, which is the optimal pH for the proteolytic enzyme pepsin. Antacids commonly used will raise the gastric pH to 4-5, which reduce pepsin proteolytic action. Stomach also secretes additional hydrochloric acid which consumes the antacid. Patient with duodenal ulcers are treated with, both $Al(OH)_3$ and $Mg(OH)_2$ mixtures (30 or 60 ml) which reduces gastric secretion.

A second potential problem of antacid is systemic alkalosis. If the antacid is water soluble or composed of readily absorbable ions, antacid may be absorbed and may exert its alkaline nature. Antacids presently in practice are water insoluble. (An exception is sodium bicarbonate which is found in most houses).

A third problem is the sodium content of antacid. Those patients who are on sodium restricted diets should be cautioned for excess of sodium when an antacid is recommended.

A fourth problem is the local effect in the G.I.T. Antacids containing calcium and aluminium salts, after being converted to soluble salts by gastric acid, tend to produce constipation, while those containing magnesium salts tend to produce laxative effect. Marketing of a constipative and laxative type of antacid in combination is quite common.

It is currently thought that antacid are effective in the reduction of pain associated with peptic ulcer, even in the absence of excess acid. As the antacids can relieve the pain of a peptic ulcer, they relatively do not have side effects when properly used, and do not prolong the ulcer.

The evaluation of antacid is difficult because the actual mechanism for their pain reduction is not known.

There are many in *vitro* techniques published. They range from adding the antacid to a given amount of hydrochloric acid and measuring the amount of acid consumed to attempt at mimicking what goes in the stomach. As the stomach is continuously emptying, the same procedures call for removing aliquots of the antacid HCl mixture from the reaction flask at given time intervals.

A plot of pH verses human pepsin stability shows pepsin to be stable up to pH 6. Above this pH, pepsin stability decreases, where complete inactivation occurs at pH 8. Even at pH 5.5, the inactivating of pepsin is only transient, with 70% of activity retained at pH 4.5.

The investigator used bovine and porcine pepsin which show maximum activity between pH 1.5-2.5, a sharp fall to 30% activity at pH-3, minimal-activity at pH-4, and zero activity at pH-5, as compared with human pepsin which has 70% activity at pH 3.5-4.5

A study shows that not only many commercial antacid preparation, but also $BaSO_4$ and a kaolin-pectin mixture (Kaopectate), inactivated pepsin by absorption of the enzyme on to the surface of particulate matter. The inactivation was independent of pH.

Vivo evaluation of antacids, is more difficult since it involves either removing sample of the gastric contents at intervals and measuring pH or introducing pH electrodes into the stomach. It can be held in a certain area and later recovered by attaching a string to it prior to swallowing of the meter. The pH changes along the individuals G.I.T. can also be followed by simply allowing the meter being carried along by bowel movement. X-Rays give the exact position of the pH meter at any given time. The pH readings are usually transmitted by radio-signal.

(C-II.1) Sodium Bicarbonate

Chemical formula : $NaHCO_3$

Molecular Weight : 84.01

Synonyms : Sodium Hydrogen Carbonate, Soda Bicarb, Baking Soda and Mitha Soda.

Standard : It contains not less than 99% and not more than 101.0% of $NaHCO_3$

Preparation

1. Solvay Process (Ammonia Soda Process) : The brine (saturated solution of sodium chloride) is saturated with ammonia to remove impurities of Mg and Fe and heated for 30 minutes. This solution is allowed to meet an ascending current of carbon dioxide in carbonating tower.

$$2NH_3 + Mg^{++} + 2\,H_2O \longrightarrow Mg(OH)_2 + NH_4^+$$

$$H_2O + CO_2 \longrightarrow H_2CO_3$$

$$NH_4 + H_2CO_3 \longrightarrow NH_4HCO_3 + H$$

$$NH_4HCO_3 + NaCl \longrightarrow NaHCO_3 + NH_4Cl$$

Brine soln.

Physical Properties

It is a white, crystalline powder or small opaque, monoclinic crystals; freely soluble in water and practically insoluble in 95% ethanol.

Chemical Properties

1. On heating it gets converted into sodium carbonate.

$$2NaHCO_3 \longrightarrow Na_2CO_3 + CO_2 + H_2O$$

2. On reaction with acids it form salt.

$$2NaHCO_3 + H_2SO_4 \longrightarrow Na_2SO_4 + 2CO_2 + 2H_2O$$

Storage : It is a hygroscopic compound and evolves carbon dioxide with moisture hence stored in air-tight containers.

Assay : It is assayed by direct acid base type of titration. Weigh accurately about 1.5 gm and dissolve in 50 ml of carbon dioxide free (freshly boiled and cooled) water and titrate with 1N hydrochloric acid using methyl orange as indicator. The standardisation of 1N HCl is carried out against primary standard anhydrous sodium bicarbonate.

$$NaHCO_3 + HCl \longrightarrow NaCl + H_2O + CO_2$$

Factor : Each ml of 1 N HCl ≡ 0.08401 gm of $NaHCO_3$

Uses : It act as antacid to neutralise excessive secretion of gastric hydrochloric acid but it produces carbon dioxide, which leads to bleaching and flatulence and also inhibits absorption of tetracycline from gastro intestinal tract. It is contra-indicated in cardiovascular patients.

It is used as wax softener in ear drop, in eye lotion for washing of eye, intravenous infusion as electrolyte replenisher and in the treatment of systemic acidosis.

Dose : 300 mg to 2 gm four times a day.

(C-II.2) Aluminium Hydroxide

Chemical Formula : $Al(OH)_3$

Molecular Weight : 77.99

Standards : It contains not less than 3.5% and not more than 4.4% w/w of Al_2O_3.

Preparation

It is prepared by the reaction of aluminium sulphate with sodium-carbonate. This aluminium carbonate is hydrolysed to give precipitate of aluminium hydroxide.

$$3Na_2CO_3 + Al_2(SO_4)_3 \longrightarrow 3Na_2SO_4 + Al_2(CO_3)_3$$

Physical Properties

It is a fine, white odourless powder or viscous-suspension. It may exhibit thixotropic properties and the liquid layer separates.

Chemical Properties

It reacts with acid to form salt.

$$Al(OH)_3 + 3HCl \longrightarrow AlCl_3 + 3H_2O$$

Aluminium hydroxide Aluminium chloride.

Purity Test

Neutralizing capacity : Dissolve 5 gm in 100 ml of water and heat with 100 ml of 0.1N HCl at 37°C. After 10,15, and 20 minutes, pH is not less than 1.8, 2.3 and 3.0 respectively.

Storage : It should be stored in air-tight containers at cool place but freezing must be avoided.

Assay : It is assayed by indirect (back) type of complexometric titration. Weigh accurately about 5 gm and dissolve in 3ml of hydrochloric acid by warming over water bath and makeup the volume to 100 ml with water. To 20 ml of this solution add 40 ml of 0.05 M disodium edetate and neutralize with 1 N NaOH using methyl red as indicator. Warm on water bath add 3 gm of hexamine and titrate with 0.05 M lead nitrate using xylenol orange as indicator.

Factor : Each ml of 0.05 M Disodium-edetate ≡ 0.002549 gm of Al_2O_3

Uses : It is a slow acting antacid with mild astringent activity causing constipation. It is used in the treatment of peptic ulcer and hyper-chlorhydria.

It is used in hypoparathyrodism and renal osteodystrophy by reducing phosphate absorption. Constipation caused by, it is relieved by magnesium salt. It reduces absorption of tetracyclin, forms complex in the G.I.T. hence pharmacologically incompatible with it.

(C-II.3) Aluminium Phosphate

Chemical Formula : $AlPO_4$

Molecular Weight : 121.95

Preparation : When aqueous solution of aluminium chloride is mixed with sodium phosphate, precipitate of $AlPO_4$ is formed.

$$AlCl_3 + Na_3PO_4 \longrightarrow AlPO_4 + 3\ NaCl$$

Aluminium phosphate.

Physical Properties

It is a white, viscous suspension; insoluble in water and readily soluble in acids. It possess thixotropic property and small amount of water is separated.

Storage : It should be stored at a temperature not exceeding 30°C but should not be freezed.

Uses : It has similar application as aluminium hydroxide but it does not interfere with absorption of phosphate salts.

Doses : 5 to 15 ml repeated as per condition of patient.

(C-II.4) Calcium Carbonate

Chemical formula : $CaCO_3$

Molecular Weight : 100.09

Synonyms : Calc carb, Calcii-carbonas, Precipitated calcium carbonate, Precipitated chalk.

Standards : It contains not less than 98.0% and not more than 100.5% calcium carbonate calculated as dried substances.

Preparation

1. When boiling solution of sodium carbonate is added to calcium chloride, a precipitate of calcium carbonate is formed. The precipitate is washed, filtered till filtrates become free from chloride ions, precipitate is dried below 105°C.

$$Na_2CO_3 + CaCl_2 \longrightarrow NaCl + CaCO_3$$

(Calcium carbonate)

2. From Lime Water : When carbon dioxide gas passes through lime water, a milky white precipitate of calcium carbonate is formed.

$$CO_2 + Ca(OH)_2 \longrightarrow CaCO_3 + H_2O$$

Calcium carbonate

Physical Properties

It is a white, odourless powder; very slightly soluble in water but solubility is increased by the presence of carbon dioxide or any ammonical salt soluble in dilute acids.

Chemical Properties

1. When excess of CO_2 is passed in the solution of calcium carbonate, soluble calcium bicarbonate is formed.

$$CaCO_3 + CO_2 + H_2O \longrightarrow Ca(HCO_3)_2$$

calcium carbonate (Insoluble) — calcium bicarbonate (soluble)

2. When it is treated with dilute acids, effervescence of carbon dioxide liberates.

$$CaCO_3 + 2\,HCl \longrightarrow CaCl_2 + H_2O + CO_2 \uparrow$$

Purity test

1. *For* Barium : To 2.6 gm of substance add 10 ml of dilute acetic acid, boil, cool and add 10 ml of calcium sulphate solution. The solution should remain clear.
2. *For* Magnesium and alkali metals : Dissolve 1 gm of substance and 10 ml of dilute hydrochloric acid and neutralize by addition of ammonia. The solution is heated to form solution of ammonium oxalates. To this add 1.5 ml of dilute sulphuric acid and evaporate to dryness on water bath, heat the residue the weight of which should not be more than 10 mg (1%).

Storage : It is stored in wellclosed containers.

Uses : It is a very good dentrifrice because of mild abrasive quality. It is also used as antacid in various preparations.

(C-II.5) Magnesium Carbonate

Chemical formula : Light — $3\ MgCO_3.Mg(OH)_2.3H_2O$; Heavy — $3\ MgCO_3.Mg(OH)_2.4H_2O$

Preparation

It is prepared from 125 parts of magnesium sulphate and 150 parts of sodium carbonate. This mixture is dissolved in boiling water for half an hour and precipitate is filtered, washed and dried in oven.

$$MgSO_4 + Na_2CO_3 \longrightarrow Na_2SO_4 + MgCO_3$$

Physical Properties

It is a white, odourless, practically insoluble in water but dissolve in acids with liberation of carbon dioxide. Heavy Magnesium carbonate occupies double volume to their weight (10 gram occupies 20 ml); while light magnesium carbonate occupies 12 times volume than their weight (10 gram occupies/20 ml).

Chemical Properties

1. Heating : Magnesium carbonate on heating forms magnesium oxide.

$$MgCO_3 \longrightarrow \underset{\text{Magnesium oxide}}{MgO} + CO_2$$

2. Treatment with acid gives magnesium chloride and carbon dioxide gas is evolved.

$$MgCO_3 + 2HCl \longrightarrow MgCl_2 + H_2O + CO_2$$

Purity test

1. **For calcium** : To 0.2 ml of ethanolic calcium standard solution (100 PPM Ca) add 1 ml of ammonium oxalate solution. After 1 minute add a mixture of 1 ml of 2 M acetic acid and

15 ml of a solution prepared by diluting 2.6 ml of solution A* to 150 ml with distilled water and shake well. After 15 minutes any opalescence produced should not be more intense than that of a standard prepared in the same manner using a mixture of 10 ml of calcium standard solution (10 PPM Ca) and 5 ml of distilled water in place of solution A (0.75%).

Solution A : Dissolve 5 gm of magnesium carbonate in 100 ml of 2M acetic acid. Boil the solution for two minutes till effervescence ceases, cool and dilute to 100 ml with 2M acetic acid.

2. **For Copper** : Dissolve 1 gm in 5 ml hydrochloric acid and 25 ml of water, boil to remove carbon dioxide, make alkaline with dilute ammonia solution; no blue colour is produced.

Storage : It is a hygroscopic in nature and should be stored in well closed containers.

Uses : It acts as antacid but liberates carbon dioxide, it acts as laxative, hence it is given in combination with aluminium or calcium salt.

Doses : 0.3 to 0.6 gm as antacid and 2-4 gm as laxative.

(C-II.6) Magnesium trisilicate

Chemical Formula : $2MgO.3SiO_2.3H_2O$ $2Mg.3SiO_2$ (anhydrous).

Molecule Weight : 314.86; 260.86 (anhydrous).

Synonyms : Epson salt, Hydrated magnesium trisilicate.

Standards : It contains not less than 29.0% Magnesium oxide and 65% SiO_2 as calculated on ignited-substances.

Preparation

It is prepared by combination of equimolar solution of sodium silicate and magnesium sulphate. It is filtered, washed and dried.

$$MgSO_4 + 2NaSiO_2 \longrightarrow MgO.SiO_2 + Na_2SO_4$$

Physical Properties

It is a fine, white powder free from grittiness, odourless, hygroscopic, practically insoluble in water and in 95% ethanol.

Storage : It is a hygroscopic substance and should be stored in well-closed containers.

Uses : It is an adsorbent and antacid having nontoxic symptoms even in large doses. It does not produce alkalosis. In presence of gastric juice it produces magnesium chloride which causes diarrhoea.

Dose : 0.5 to 2 gm every 6 hourly.

(C-II.7) Magnesium Oxide

Chemical Formula : MgO

Molecule Weight : 40.30

Synnonym : Magnesia.

Preparation

It is prepared by heating magnesium carbonate till it becomes dull red or until carbon dioxide is no longer evolves.

$$MgCO_3 \longrightarrow MgO + CO_2$$

Physical Properties

It is a an odourless, white amorphous powder; very slightly soluble in water, practically insoluble in alcohol and soluble in dilute acids. The light magnesium oxide occupies 10 times volume as comparative to their weight while heavy magnesium oxide occupies twice the volume to their weight.

Chemical Properties

1. *With moisture* : It reacts with atmospheric moisture and forms magnesium hydroxide.

$$MgO + H_2O \longrightarrow Mg(OH)_2$$

Magnesium hydroxide.

2. *With air* : It reacts with atmospheric carbon dioxide and form magnesium carbonate.

$$MgO + CO_2 \longrightarrow MgCO_3$$

Magnesium carbonate.

Storage : Stored in tightly closed containers.

Uses : It has similar action as that of magnesium trisilicate.

C (III) Protectives and adsorbents

This group of gastro-intestinal agents are used for the treatment of mild diarrhoea. Diarrhoea is a serious condition with loss of fluids and electrolytes which can lead to dehydration and electrolyte imbalance. Diarrhoea may be acute or chronic. Acute diarrhoea is caused by bacterial-toxins, chemical poisons, drug, allergy or with other disease causing tissue damage or irritation in intestinal tract and change in osmotic load. Chronic diarrhoea may be caused by gastro intestinal surgery, chronic inflammatory condition, carcinomas and other defects of absorption. The anti-diarrhoeal agents may act as.

1. Anti-bacterial
2. Anti-spasmodic or anti-motility agents and
3. Adsorbent-protective

The adsorbent-protectives are supposed to adsorb toxins, bacteria and viruses and provides a protective coating on the intestinal mucosa. Bismuth salts, special clays and activated charcol are generally act as adsorbent-protectives.

C (III-1) Bismuth Subcarbonate

Chemical Formula : $(BiO)_2\ CO_3.\ 1/2\ H_2O$.

Molecular Weight : 529.00

Standards : It contains not less than 80% and not more than 82.5% of bismuth as dried substance.

Preparation

The metallic bismuth is treated with 50% nitric acid and solution is heated, Bismuth nitrate is formed and, to this solution sodium carbonate is added in cold condition, precipitate is filtered, washed and dried at 55°C.

$$2Bi + 8HNO_3 \longrightarrow 2Bi(NO_3)_3 + 2\ NO + 4H_2O$$

$$2Bi(NO_3)_3 + 6\ Na_2CO_3 + H_2O \longrightarrow [(BiO)_2\ CO_3].\ H_2O + 12\ NaNO_3 + 4\ CO_2.$$

Physical Properties

It is a white, odourless; practically insoluble in water, 95% ethanol and ether; but soluble in mineral acids with effervescence.

Purity test

1. *For Copper* : To 5 ml of 5% solution in 20% nitric acid, add 2 ml of 10 M NH_3, dilute to 50 ml with water and filter. To 10 ml of the filtrate add 1 ml. of a 0.1% w/v solution of sodium diethyl dithiocarbonate. Any colour produced is not more intense than that produced by treating at the same time and in the same manner with a solution containing 0.25 ml of copper standard solution (10 PPM Cu) diluted to 10 ml with water (50 PPM).

2. *For silver* : To 2.0 gm add 1 ml of water and 4 ml of nitric acid, Heat gently to dissolve and ditute to 11 ml with water. Cool, add 2 ml of 1 M HCl and allow to stand for 5 minutes protected from light. Any opalescence produced should not be more intense than that obtained by treated at the same time and in the same manner with a mixture of 10 ml of silver standard solution (5 PPM Ag), 2 ml of 1 M HCl and 1 ml of HNO_3 (25 PPM).

Storage : It is stored in well closed, light resistant containers.

Uses : It is used internally in systemic toxicity gastric disorders, ulcers, diarrhoea, dysentery, ulcerative-colitis etc. as protective.

C (III-2) Light Kaolin

Chemical Formula : $Al_2O_3.2SiO_2.\ 2H_2O$.

Molecular Weight : 258

Synonyms : The china clay, Hydrated aluminium silicate.

Preparation

The china clay (hydrated aluminium silicate) is finely powdered and made free from gritty particles. This powder is elutriated to remove other soluble impurities and dried.

Physical Properties

It is a white powder, tasteless, practically insoluble in water and in mineral acids.

Storage : It is stored in well closed containers.

Uses : Light kaolin is used as an adsorbent. It is administered by mouth in the symptomatic treatment of enteritis, colitis and dysentery associated with food poisoning and alkaloidal poisoning. It may be administered as suspension in water.

Light Kaloin is applied externally as dusting powder either undiluted or mixed with other protectives. It is an ingredient of toilet powders and a diluent of disinfactant powders. It is also used for filteration purposes.

Dose : 15 to 75 gram in three divided dose daily.

C [IV] Saline Cathartics

Saline Cathartics (Purgatives) are agents which promote evacuation of faecal matter from the bowels. Laxative are mild cathartics. Cathartics are used to.

1. Ease defecation in patients with painful haemorrhoids or other rectal disorders.
2. To avoid potentially hazardous rise in blood pressure during defecation in the patients of hypertension, cerebral, coronary, or other arterial disease.
3. Relieve acute constipation.
4. Remove solid material from intestinal tract prior to X-Ray.

Laxatives are propelled through the intestinal tract by wave like contraction called peristaltic movement. A laxative habit may develop in those patients who keeps rhythm of defecation through laxatives. These agents are of four types.

1. *Stimulants* : They act by local irritation on the intestinal tract and increase peristaltic activity. e.g. Phenopthelein, aloin, cascara, rhubarb, senna, bisacodyl.
2. *Bulk forming* : These are made of cellulose or non-digestable polysaccharides. They get swells with water and increase bulk, thereby stimulate, peristalsis. e.g. Psyllium seed, Methyl cellulose, Sodium carboxy methyl cellulose, Kararga gram.

3. *Emollient laxative* : They act either as lubricant and facilitating compact faecal material to go out or act as stool softeners. e.g. Mineral oil, deoctyl sodium sulpho-succinate etc.

4. *Saline cathartics* : These agents increases osmotic load of gastro intestinal tract. These salts are poorly absorbable anions and cations. The body relieves the hyper-tonicity of the gut by secreting additional fluid into the intestinal-tract. These agents are water soluble taken with large amount of water and prevent excessive loss of body fluid.

C(IV-1) Sodium Potassium Tartarate

Chemical Formula : CH(OH)COONa

$$CH(OH)COOK \cdot 4H_2O$$

Molecular Weight : 283.23

Synonyms : Rochelle salt, Seignette's salt.

Preparation

It is prepared by heating sodium carbonate and potasium bi-tartarate at 60°C till no more carbon dioxide evolves. The solution is cooled, filtered, and evaporated to crystallization.

$$\underset{}{Na_2CO_3} + \underset{\text{Pot. bitartarate}}{2C_4H_4O_6KH} \longrightarrow \underset{\text{Sod.Pot.tartarate}}{2C_4H_4O_6NaK} + H_2O + CO_2$$

Physical Properties

It is an odourless, colourless or white, crystalline powder, soluble in water but very slightly soluble in alcohol.

Chemical Properties

When it is ignited above its M.P. (74°C), it becomes red and forms bicarbonates of sodium and potassium.

$$2C_4H_4O_6NaK + 5O_2 \longrightarrow K_2CO_3 + Na_2CO_3 + 4H_2O + 6CO_2$$

Uses : It is a saline cathartic, diuretic and reduces acidity of urine. They cause watery evacuation of bowel after 1 to 2 hour with out producing irritation.

Dose : 8 to 16 gm.

C (IV-2) Magnesium Sulphate

Chemical Formula : $MgSO_4 \cdot 7H_2O$

Molecular Weight : 246.47

Synonym : Epsom salt.

Standards : It contains not less than 99.0% $MgSO_4$ and not more than 100.5% of Mg SO_4 calculated as dried substance.

Preparation

1. When dolomite (Mg CO_3 . $CaCO_3$) is treated with sulphuric acid it forms water soluble magnesium sulphate and sparingly soluble calium sulphate.

$$MgCO_3 \bullet CaCO_3 + 2H_2SO_4 \longrightarrow MgSO_4 + CaSO_4 + 2H_2O + 2CO_2$$

2. When magnesium carbonate is neutralized with hot dilute sulphuric acid, it forms magnesium sulphate.

$$MgCO_3 + H_2SO_4 \longrightarrow MgSO_4 + H_2O + CO_2$$

Physical Properties

It is a white crystalline powder, odourless, highly soluble in boiling water, freely soluble in water and practically insoluble in 95% ethanol.

Chemical Properties

1. When it is heated on 200°C it looses water of crystallisation.

$$MgSO_4 \bullet 7H_2O \xrightarrow[200°C]{} MgSO_4 + 7H_2O$$

2. When anhydrous compound is heated above 200°C it liberates sulphur dioxide and oxygen.

$$MgSO_4 \longrightarrow MgO + SO_2 + O_2$$

Incompatibility : It is incompatible with alkali carbonates and bicarbonates; and with potassium and ammonium bromide.

Storage: It should be stored in air tight containers in a cool place.

Uses : It acts as saline purgative. It reduces the normal absorption of water from the intestine, with the result that the bulky fluid contents distant the bowl, active reflex peristalsis is exited and evacuation of intestinal content take place.

The intravenous injection is used to lower intra-cranial pressure and to control the convulsion in acute uraemia and eclampsia. Wet dressings of a 25% solution of magnesium sulphate are used in the treatment of carbuncles and boils. It is also used in the treatment of hypo magnesia in the cattle.

Dose : 10 to 30 gm daily.

(D) TOPICAL AGENTS

Topical means "use of compounds" on the skin or mucous membrane or in the the body cavities (*Example* : Oral, vaginal and colonic cavities). These agents are absorbed into the circulatory system and distributed in various organs and tissues. These agents are commonly applied at skin, eye, nose, ear, throat, rectum, vagina and urethra as powders, ointments, creams, paste, lotion, spray or transdermal patches. These routes are advantagious over systemic route of administration because it by-pass burden of metabolism. It is beneficial in penetration of antiseptic to prevent deep infection and prevents the body from toxic effects of antiseptic agents. The rate of absorption depends on thickness of outer skin and blood circulation at the site of application.

These agents are broadly classified according to their action.

A. PROTECTIVE AND ADSORBENTS

B. ANTI-MICROBIAL AGENTS

C. ASTRINGENTS

D. MISCELLANEOUS

(D-a) PROTECTIVES

Protectives are compounds, applied on the skin to protect, ulcer or open wounds from irritation. The ideal characters of protective agents are: (1) insoluble in water; (2) chemically inert; (3) biologically inert and (4) available as fine particles. The insolubility of particles are desirable to limit the absorption of the compounds through skin, difficult to wash and diminishes metallic properties on tissue. The chemical inertness of substances are required to prevent any local irritation or allergic reactions. The small particle size is required to increase surface area and allowing them to adsorb the moisture present on the skin more effeciently.

These agents are applied on the area of constant irritation due to moisture, friction, inflammed due to friction, allergy, etc. These are mixed with the exudated fluid and helps in drying of crust. The following compounds are used as protective.

(a-1) Talc

Chemical formula : $3MgO \bullet 4SiO_2 \bullet H_2O$

Molecular Weight : 278.00

Synonyms : Purified Talc, Talcum, French chalk, Soap stone.

Standards : It is a powdered, purified, natural, hydrated magnesium silicate containing small fraction of aluminium silicate.

Preparation

It is a mineral obtained in the Araveli hills of Udaipur region in Rajasthan. The mineral is powdered and treated with dilute hydrochloric acid. The supernated liquid is washed and removed till it becomes free from acid. The precipitate is dried and powdered.

Physical Properties : It is a very fine white powder, free form grittiness, odourless, tasteless, readily adhere to the skin and practically insoluble in dilute acids or alkalis.

Storage : It is stored in well-closed containers.

Uses : It is a dusting powder, medicated with zinc oxide or and boric acid. It should not be applied over broken skin, wounds, surgical incision and surgical gloves because it produce granulomata. It is good filtering media because it does not absorb or adsorb any principle active substances. It is also used as excipient and as lubricant in the preparation of pills and tablets.

(a-2) Zinc Oxide

Synonym : Philosopheriwool

Chemical formula : ZnO

Molecular Weight : 81.38

Standard : It contain not less than 99.00% as calculated with reference to ignited zinc oxide.

Preparations

1. It is manufactured by heating metallic zinc in current of air. The vapour of zinc burns to form the oxide which is collected as fine powder.

$$2\,Zn + O_2 \longrightarrow 2\,ZnO$$

2. The pure form of zinc oxide is prepared by reaction of zinc sulphate and sodium carbonate, precipitate of zinc carbonate is obtained. The precipitate is washed, dried and gently ignited to form zinc oxide.

$$ZnSO_4 + Na_2CO_3 \longrightarrow ZnCO_3 + Na_2SO_4$$

$$ZnCO_3 \longrightarrow ZnO + CO_2$$

Physical Properties

It is a white or yellowish white soft powder free from grittiness, odourless, tasteless, practically insoluble in water and soluble in dilute mineral acids.

Chemical Properties

1. On heating with hydrogen or carbon above 450°C, metallic zinc is formed.

$$ZnO + H_2 \longrightarrow Zn + H_2O$$

$$ZnO + C \longrightarrow Zn + CO_2$$

2. On heating with cobalt nitrate solution, it form green mass of cobalt zincate

$$Co(NO_3)_2 \longrightarrow 2CoO + HNO_2 + O_2$$

$$ZnO + CoO \longrightarrow ZnCoO \text{ or } CoZnO_2$$

Storage : It is stored in well closed containers. It gradually absorbs moisture and carbor dioxide from atmosphere.

Use : It is a mild antiseptic and astringent, used in pastes, lotions and dusting powders. It acts as soothing and protecting agent in eczema and as a protective for slight excoriations (removal of skin).

(a-3) Calamine

Chemical formula : $ZnO \bullet Fe_2O_3$

Standards : It contains not less than 98% and not more than 100.5% of zinc oxide calculated as ignited substance.

Preparation

The naturally occuring ore zinc carbonate is calcinated. The calcinated product is powdered and passed through a 100 mesh sieve to obtain fine powder possesing good cohesive and adhering property on skin. The fine powder obtain in zinc oxide is mixed with small proportion of ferric oxide (1%) which imparts pink colour to powder.

Physical Properties

It is a fine amorphous pink or reddish powder, odourless and almost completely soluble in hydrochloric acid with effervescence.

Purity tests

1. *Water soluble dyes* : A 10% solution in water is filtered, the filtrate should be colourless.
2. *Alcohol soluble dyes* : A 10% alcoholic solution in (90%) ethanol is filtered, filtrate should be colourless.
3. *Alkaline substances* : Digest 1 gram substance with 20ml warm water. It is neutralized with 0.1 N sulphuric acid using phenolphthlein as indicator. Not more than 0.2 ml of acid should consume.

Storage : It is stored in well-closed containers.

Uses : Calamine is used as dusting powders, lotions, and ointments applied to the skin for soothing, adsorbents and protective properties. This is more acceptable than zinc oxide. It also has a problem where exudation of fluids take place. It is a mild astringent, its lotion cools the skin by evaporation and is useful for relieving the pain and swelling of sun burn. Oily lotion is used in the treatment of eczema as soothing agents. The phenolic lotion contains 1% liquified phenol which provides a local anaesthetic and antipuritic (anti-itching) action.

(a-4) Zinc Stearate

Chemical Formula : $(C_{17}H_{35}COO)_2Zn$.

Molecular Weight : 631

Standards : It contains not less than 12.5% and not more than 14.5% of zinc oxide.

Preparation

The stearic acid is first treated with a hot solution of sodium hydroxide or sodium carbonate and forms sodium stearate which is treated with zinc sulphate. A precipitate of zinc stearate is formed.

Chemical Properties

When it is treated with alkalis or zinc salts it forms sodium or zinc stearate.

$$NaOH + C_{17}H_{35}COOH \longrightarrow C_{17}H_{35}COONa + H_2O$$

$$Na_2CO_3 + 2C_{17}H_{35}COOH \longrightarrow 2C_{17}H_{35}COONa + H_2O + CO_2$$

$$2C_{17}H_{35}COONa + ZnSO_4 \longrightarrow (C_{17}H_{35}COO)_2Zn + Na_2SO_4$$

Sod.stearate zinc sulphate zinc stearate

Purity tests

1. *Alkali and alkaline earths* : To 1 gm of substance add 25 ml of water and 5 ml of hydrochloric acid, boil, filter and wash with hot water. Add dilute ammonia solution to make the filtrate alkaline and add ammonium sulphide solution in excess to precipitate zinc as zinc hydroxide. Filter, add 0.5 ml of sulphuric acid to filtrate, evaporate to dryness and ignite in a tared crucible, residue should not be more than 20 mg.
2. *Free fatty acid* : To 5 gm of substance add 100 ml of solvent ether and shake for half an hour, filter and evaporate 50 ml of filtrate to dryness. It should not be more than 50 mg.

Uses : It is used as mild-astringent, antimicrobial and protective. It has advantages over talc that, it did not wet by moisture, hence more desirable for skin problem where large amount of fluid is exuded. Its dust causes pulmonary inflammation when used as dusting powder, so it is not suitable for infants and children.

(a-5) Titanium Dioxide

Chemical Formula : TiO_2

Molecular Weight : 79.90.

Standards : It contains not less than 98% of titanium with reference to dried substances.

Physical Properties

It is a white amorphous infusible powder; odourless; tasteless; practically insoluble in water, and in dilute mineral acid but soluble in hydrofluoric acid and in hot sulphuric acid.

Chemical Properties

1. On heating with carbon in a current of chlorine,metal chloride is formed.

$$TiO_2 + C + Cl_2 \longrightarrow TiCl_4 + CO_2$$

2. On heating with carbon or aluminium metal

$$TiO_2 + C \longrightarrow Ti + CO_2$$

$$3\ TiO_2 + 4\ Al \longrightarrow 3Ti + 2\ Al_2O_3$$

Storage : It is stored in well-closed containers. It should not be stored or allowed to come in contact with aluminium.

Uses : It is a opacifying agent which reflects ultraviolet radiation due to high refrective index (2.7). It is also called as "Solar ray protective". It is used as a white pigments in paints and cosmetic.

(a-6) Silicon Polymers

Synonym : Dimethicone, Dimethylpolysiloxane, Simethicone, Silicon oil.

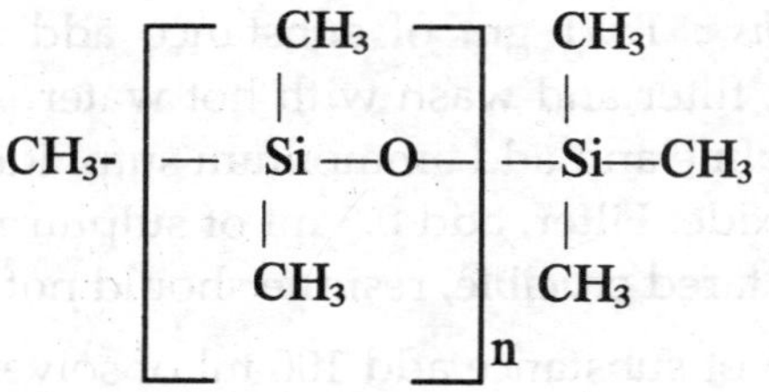

Preparation

The Dimethicone is prepared by the polymerisation of dimethyl silicone ether.

Physical Properties

It is a clear colourless, odourless liquid. It is soluble in ether, Xylene, chlorinated hydrocarbons and solvent naphtha. Its various grades like dimethicone 20, 200, 350 and 500 are soluble in amyl acetate, cyclohexane, light petroleum and kerosene. They are practically insoluble in water and acetone. The various grades are as per their viscosity.

Chemcial Properties

These are chemically innert but undergoes polymerisation similar to carbon.

Storage : It should be stored in light resistant containers.

Uses : It is hydrophobic (water repellent) liquid having low surface tension, thermostable and resistant to most chemical substances but affected with strong acids.

It is used as creams or lotions for protecting skin from irritating substances, trauma from urine or faecal discharge and also for preventing bedsores.

Its water repellent property is used over the surface of vials and syringe. A 2% of silicon oil "1000" in xylene solution is filled, excess is drainout and dried at 110°C; the glass is then heated at 250 - 270°C so that silicon oil forms bond with glass. It is also used in the treatment of flatulence and as **anti-foaming agent**.

(D-B) ANTI-MICROBIALS AND ASTRINGENTS

Astringents and anti-microbial agents act externally on microbes to prevent the growth or to kill them. Various terms used as anti-microbial agents are :

Antiseptic

These agents kill or inhibits the growth of pathogenic micro-organisms, i.e. bacteria, fungi, protozoas, etc; over living tissues.

Disinfectant

These agents also kill or inhibit the growth of micro-organism over inanimate objects (non-living substrate like equipments, rooms, instruments, tables etc.). These agents are more potent than antiseptic.

Germicide

Germicide are the agents which kill germs. Germ is a wide term. Specific terms like bacteria, fungi or virus are used more than agents which can kill these pathogens are called as bacteriocidal, fungicidal and virucial respectively.

Bactericidal

These agents kill the bacteria are called bactericidal.

Bacteriostatic : Those agent which inhibits the growth of bacteria or arrest the growth but does not kill the bacteria are called bacteriostatic.

Sterilization : It is a process by which all the micro-organisms and their spores are either killed or removed and resultant product will be free from living organism. It can be done by chemical, mechanical (dry or moist heat) or by radiation.

Mechanism of action : The anti-microbial agents discussed in this chapter have the following type of mechanism of action.

1. Oxidation
2. Halogenation
3. Protein-precipitation.

1. *Oxidation :* The highly oxidising agents like H_2O_2, $KMnO_4$ etc. act as anti-microbial agents. These agents react with reducing groups present in the proteins. They changes the behaviour of protein or alter the confirmation (shape) of proteins and inhibits the growth of micro-organism e.g., the sulphur containing proteins (Proteins containing cystin, cysteine and methionine aminoacids contains sulphur) act as reducing group. These SH group get converted into di-sulphide linkage (-S-S-) due to oxidising agents.

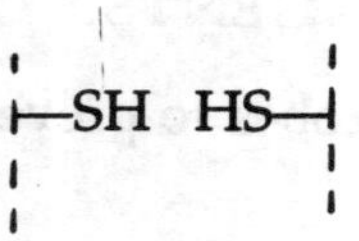

Proteins containing cystin, cysteine and methionine type amino-acid after oxidation.

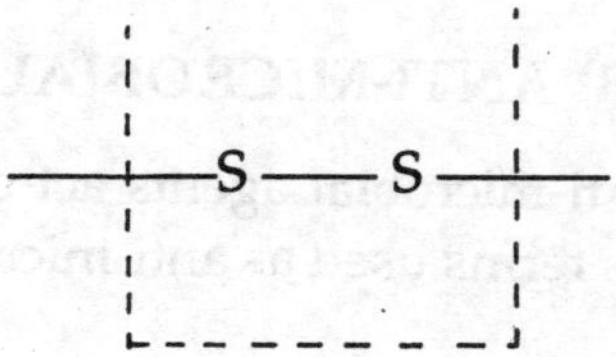

Proteins containing cystin, cysteine and methionine type amino-acid. (Prior to oxidation).

2. *Halogenation* : The halides and hypohalides are the main causative group which can change the peptide linkages by haloginating secondary amino group present in peptide linkages.

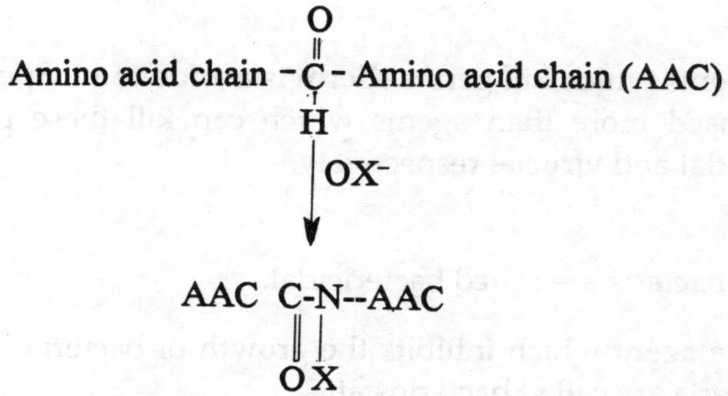

These type of changes in protein alter the functions of protein and forces (Hydrogen-bonding) responsible for proper confirmation of protein molecule. This results in destruction of protein.

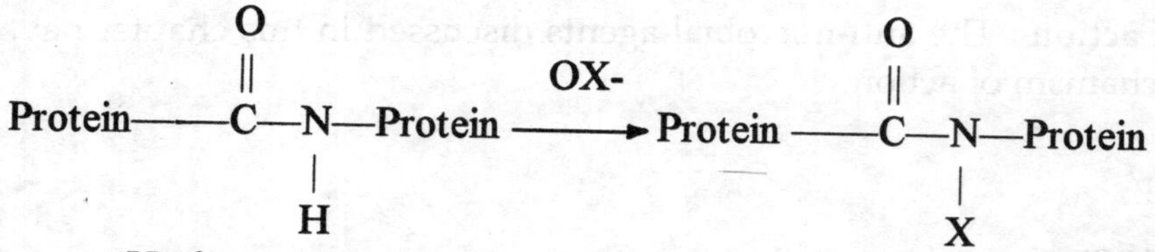

Hydrogen bond present

Hydrogen bond not possible.

3. *Precipitation of proteins* : The metals of group I(B), II(B), and III(A) like Ag(I), Cu(II), Zn(II), Al (III), etc. are having large charge/radius ratio or has strong electrostatic field hence these metal ions form a complex in which polar group of protein act as ligands. It changes the properties of protein or causes precipitation. Certain metals (Hg, As, Sb) are enzyme specific and forms covalent bond with enzymes leads to destruction of enzymes.

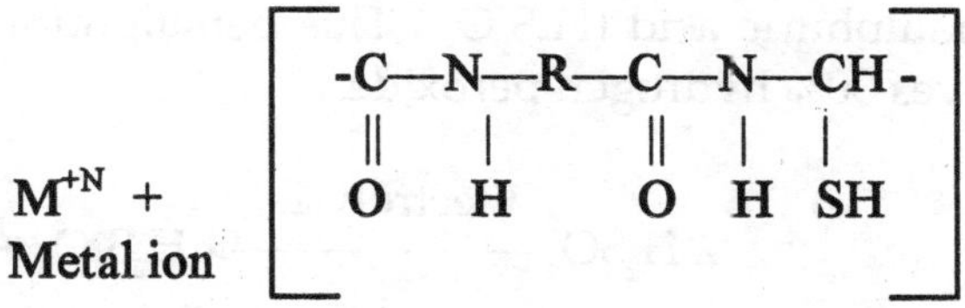

Protein (ligand)

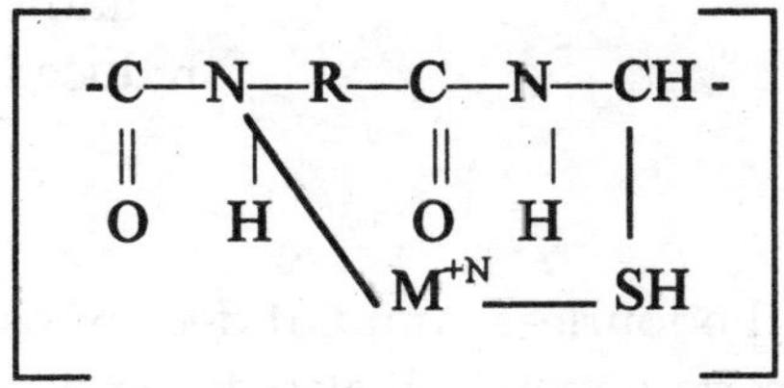

Metal - Protein Complex (Chelate).

Astringents : Astringents are agents which precipitate local or surface protein. It is a surface phenomena and does not affect host tissue. These agents cause shrinkage of tissue cells, lead to construction of capillaries and small blood vessels. These properties are used to stop bleeding from small cuts and also as anti-perspirants in deodarants because of their ability to constrict pores and destroy micro-organisms producing unpleasent odour.

Astringents are applied over wounds in small quantity to stimulate the growth of new tissue but in higher concentration it produces irritation or corrosive effect. The corrosive property is useful in removal of undesirable tissue e.g. warts.

(b.1) Hydrogen Peroxides

Chemical Formula : H_2O_2

Molecular Weight : 34.02

Standards : Hydrogen peroxide solution (100 ml) contains not less than 26% w/w and not more than 28% w/w of H_2O_2, corresponding to 100 times, its volume of available oxygen.

Preparation

1. *From Barium Peroxide* : A thick paste of barium peroxide is placed in ice cold water and a calculated quantity of dilute-sulphuric acid (cooled in ice) or phosphoric acid in presence of carbon dioxide are mixed at lower temperature. The obtained hydrogen peroxide is 10-20% w/w.

$$BaO_2 + H_2SO_4 \longrightarrow BaSO_4 + H_2O_2$$

$$3\,BaO_2 + 2\,H_3PO_4 \longrightarrow Ba_3(PO_4)_2 + 3H_2O_2$$

2. *Electrolysis of sulphuric acid* : A 50% ice cold solution of sulphuric acid is electrolysed, which produces persulphuric acid ($H_2S_2O_8$). This persulphuric acid is distilled under reduced pressure gives 30% hydrogen peroxide.

$$2\,H_2SO_4 \xrightarrow{\text{Electrolysis}} \underset{\text{Persulphuric acid}}{H_2S_2O_8} + H_2$$

$$H_2S_2O_8 + 2H_2O \xrightarrow[\text{under- reduced pressure}]{\text{distilled}} 2\,H_2SO_4 + \underset{\text{Hydrogen Peroxide}}{H_2O_2}$$

Physical Properties

It is a clear, colourless and odourless liquid. It decomposes in contact with oxidising agents, organic compounds and certain metals. It also decomposes with increasing temperature.

Chemical Properties

The hydrogen peroxide react as either an oxidising or reducing agent depending upon chemical envoirnment.

Reduction : Reduction reaction take place in acidic medium but usually involves metallic oxides.

a. $Ag_2O + H_2O_2 \longrightarrow 2Ag + H_2O + O_2$

b. $2\,KMnO_4 + 3H_2SO_4 \longrightarrow K_2SO_4 + 2\,MnSO_4 + 3H_2O + 5\,[O]$

$[\,[O] + H_2O_2 \longrightarrow H_2O + O_2] \quad X\,5$

$2\,KMnO_4 + 3\,H_2SO_4 + 5H_2O_2 \longrightarrow K_2SO_4 + 2\,MnSO_4 + 8\,H_2O + 5O_2.$

c. With Manganese dioxide and red lead oxide in acidic medium

$$MnO_2 + H_2SO_4 + H_2O_2 \longrightarrow MnSO_4 + 2H_2O + O_2$$

$$Pb_3O_4 + 6HNO_3 + H_2O_2 \longrightarrow 3Pb(NO_3)_2 + 4H_2O + O_2$$

d. With ozone

$$H_2O_2 + O_3 \longrightarrow H_2O + 2O_2$$

e. With Potassium ferricynide in alkaline medium

$$2\,K_3[Fe(CN)_6] + 2KOH + H_2O_2 \longrightarrow 2K_4[Fe(CN)_6] + 2H_2O + O_2$$

f. With chlorine and bromine

$$I_2 + H_2O_2 \longrightarrow 2HI + O_2$$

g. With Hyrohalides

$$NaOCl + H_2O_2 \longrightarrow NaCl + H_2O + O_2$$

$$NaOBr + H_2O_2 \longrightarrow NaBr + H_2O + O_2$$

Oxidation : When oxidising agent reacts, it forms two molecule of oxide ions (O^{-2}).

$$O_2 \longrightarrow 2O^-$$

a. This type of reaction occurs in acidic medium and produces alkali.

$$H_2O_2 + 2KI \longrightarrow 2KOH + I_2$$

b. With black lead sulphide to form white lead sulphate

$$PbS + 4H_2O_2 \longrightarrow PbSO_4 + 4H_2O$$

c. With ferrous sulphate

$$2FeSO_4 + H_2O_2 + H_2SO_4 \longrightarrow Fe_2(SO_4)_3 + 2\,H_2O$$

d. With sodium sulphite it forms sulphate

$$Na_2SO_3 + H_2O_2 \longrightarrow Na_2SO_4 + H_2O$$

e. With nitrites it forms nitrates

$$KNO_2 + H_2O_2 \longrightarrow KNO_3 + H_2O$$

f. With arsenites it forms arsenates

$$Na_3\,AsO_3 + H_2O_2 \longrightarrow Na_4\,AsO_4 + H_2O$$

g. With pot. ferrocynate it forms pot fericynate

$$2K_4Fe(CN)_6 + H_2O_2 \longrightarrow 2K_3Fe(CN)_6 + 2\,KOH$$

h. With formaldehyde it forms formic acid

$$HCHO + H_2O_2 \longrightarrow HCOOH + H_2O$$

Purity test

1. *Acidity* : 10 ml of Hydrogen peroxide should not require more than 0.5 ml of 0.1 M NaOH.
2. *Organic stabilizers* : Extract 20 ml of hydrogen peroxide with successive quantity of chloroform, evaporate chloroform-extract at reduced pressure in dry desiccator and temperature should not exceed 25^{0}C. This residue should not be more than 10 mg (500 PPm).

Incompatibility

It is incompatible with most organic substances and with alkalies, iodides, permagnates, and other oxidizable substances.

Storage : It should be stored in light resistant (black) containers in cool place (not exceeding 50°C). The bottles should be closed with glass or suitable plastic caps.

Assay : It is assayed by direct redox (permagnometric) type of self-indicating titration. Dilute 1 gm to 100 ml with water. To 10 ml of dilute solution, add 20 ml of 2 N sulphuric acid and titrate with 0.1 N potassium permagnate till pink colour persist. The potassium permagnate is standardize against anhydrous sodium oxalate as primary standard.

Factor : Each ml of 0.1 N$KMnO_4$ ≡ 0.0017 gm of H_2O_2 or 0.56 ml oxygen.

$$2KMnO_4 + 3\,H_2SO_4 \longrightarrow K_2SO_4 + 2MnSO_4 + 3\,H_2O + 5[O]$$

$$[\,H_2O_2 + [O] \longrightarrow H_2O + O_2\,] \times 5$$

$$2\,KMnO_4 + 3\,H_2SO_4 + 5H_2O_2 \longrightarrow K_2SO_4 + 8\,H_2O + 5\,O_2 + 2MnSO_4$$

Uses : It has antiseptic deodorant and bleaching action. It evolves oxygen rapidly and achieves antiseptic action thus does not combine with proteins of skin surface. It is useful for removal of dead tissue and bacterial nests from dirty-wounds and, removing adherent blood stained dressings, removal of wax from ear and deodorant gargles or mouth washes.

(b-2) Potassium Permanganate

Chemical formula : $KMnO_4$

Molecular Weight : 158.03

Standards : It contains not less than 99% and not more than 100.5% of $KMnO_4$.

Preparation

It is prepared by heating a mixture of potassium hydroxide with manganese dioxide, which produces a green mass of potassium manganate. This green mass is extracted with boiling water with a current of chlorine or carbon dioxide. The solution is evaporated to crystallization.

$$6KOH + 3MnO_2 + KClO_3 \longrightarrow 3K_2MnO_4 + KCl + 3H_2O$$

$$2K_2MnO_4 + 2Cl_2 \longrightarrow 2KCl + 2\,KMnO_4$$

$$3K_2MnO_4 + 2CO_2 \longrightarrow 2KMnO_4 + MnO_2 + 2K_2CO_3$$

Physical Properties

It is a dark purple or brownish black granules; cylindric or prismatic crystal having metallic lusture; odourless; soluble in cold water, freely soluble in boiling water and it decomposes on contact with certain organic substances.

Chemcial Properties

1. Neutral or alkaline solution produces brown precipitate of manganese dioxide.

$$2KMnO_4 + H_2O \longrightarrow \underset{\text{(brown)}}{2\,MnO_2} + 2\,KOH + 3\,[O]$$

2. In neutral or alkaline solution, iodides are oxidized to iodates.

$$2KMnO_4 + H_2O + KI \longrightarrow 2MnO_2 + 2KOH + KIO_3$$

3. Potassium permanganate in acidic condition librates halogen from halide salts.

$$2\,KMnO_4 + 10KI + 8H_2SO_4 \longrightarrow 6K_2SO_4 + 2MnSO_4 + 5I_2 + 8H_2O$$

4. It reacts with ferrous sulphate in acidic medium to form ferric sulphate.

$$2\,FeSO_4 + 2KMnO_4 + 4H_2SO_4 \longrightarrow K_2SO_4 + 2MnSO_4 + Fe_2(SO_4)_3 + 4H_2O$$

Purity test

1. *For chloride and sulphate limit test :* Its solution is purple in colour. The colour of solution interfere with limit test of chloride and sulphate. The purple colour is destroyed by boiling its solution with 95% alcohol and precipitate of manganese dioxide is removed. After filtration usual test is carried out.

$$2KMnO_4 + 3C_2H_5OH \longrightarrow 2MnO_2 + 2KOH + 2H_2O + 3CH_3CHO$$

Incompatibility

It is incompatible with reducing agents, iodides, most organic matter's, and its acidic solution is incompatible with hydrogen peroxide.

Storage : It should be stored in air tight containers to avoid contact with organic matter.

Assay : It is assayed by direct iodometric type of redox titration. The 0.3 gm of substance dissolved to produce 100 ml with water. Take 20 ml of solution; add 1 gm of KI and 10 ml of 2N HCl and, titrate the liberated iodine with 0.1 M $Na_2S_2O_3$ using starch as indicator.

Factor : Each ml 0.1M $Na_2S_2O_3 \equiv 0.003160$ gm of $KMnO_4$

$$2KMnO_4 + 10\,KI + 8\,H_2SO_4 \longrightarrow 6\,K_2SO_4 + 2MnSO_4 + 5I_2 + 8\,H_2O$$

$$2Na_2S_2O_3 + I_2 \longrightarrow 2NaI + Na_2S_4O_6$$

Uses : It is a strong oxidising agent and is used for disinfactant, deodorant, cleaning of ulcers or abscesses, gargles, mouthwashes, vaginal irrigation, urethral irrigation and, first-aid treatment in snake bite, etc.

(b-3) Calcium Oxychloride

Chemcial formula : Ca(OCl)Cl

Molecular Weight : 136.98

Synonyms : Bleaching powder, Chlorinated lime, Chloride of lime.

Standards : It should contain not less than 30% w/w of available chlorine.

Preparation

It is prepared by passing chlorine gas over dry calcium hydroxide (slaked lime) in lead chamber for 18-24 hours. In this process slaked lime gets saturated with chlorine gas.

$$Ca(OH)_2 + Cl_2 \longrightarrow Ca(OCl)Cl + H_2O$$

Physical Properties

It is a dry, dull white powder with characteristic odour. It absorbs moisture on exposure to air and gradually decomposes.

Chemical Properties

1. On exposure to air it decomposes to calcium carbonate and calcium chloride

   $2CaOCl_2 + CO_2 + H_2O \longrightarrow CaCO_3 + CaCl_2 + 2HOCl$

2. Decomposition

   a. On prolong storage it decomposes to $Ca(ClO_3)_2$

   $2CaOCl_2 \longrightarrow Ca(ClO_3)_2 + 5CaCl_2$

   b. In presence of cobalt chloride

   $6CaOCl \xrightarrow{CoCl_2} 2CaCl_2 + O_2$

   c. In solution

   $2CaOCl_2 \longrightarrow Ca(OCl)_2 + CaCl_2$

   d. In presence of dil acid

   $CaOCl_2 + 2HCl \longrightarrow CaCl_2 + H_2O + Cl_2$

3. Reaction with acetone

   $3CaOCl_2 + 2CH_3COCH_3 \longrightarrow 2CHCl_3 + (CH_3COO)_2Ca + 2Ca(OH)_2$

4. Reaction with alcohol

   $4CaOCl_2 + 2C_2H_5OH \longrightarrow 2CHCl_3 + (HCOO)_2\,Ca + Ca(OH)_2 + 2H_2O + CaCl_2$

5. Reaction with NH_4OH

   $3CaOCl_2 + 2NH_4OH \longrightarrow CaCl_2 + 5H_2O + N_2$

Storage : It should be stored in well-closed container. It slowly decomposes with loss of chlorine.

Assay : It is a direct iodometric redox type of titration. The weighed quantity is taken and 1 gm of KI is added. To this mixture 10ml of acetic acid is added. The mixture is kept aside, in dark for 10 mintues. It is titrated against N/10 $Na_2S_2O_3$ using starch as indicator.

$$CaOCl_2 + 2CH_3COOH \longrightarrow (CH_3COO)_2Ca + HOCl + HCl$$

$$HOCl + HCl \longleftrightarrow Cl_2 + H_2O$$

$$2KI + Cl_2 \longrightarrow 2\,KCl + I_2$$

$$2Na_2S_2O_3 + I_2 \longrightarrow Na_2S_4O_6 + 2NaI$$

Uses : It is used as a bleaching agent, to decolorise most of dyes. It is used in the preparation of detergents. It is used as a disinfectant in wounds, or sterile rooms; and swimming pools. It is used in water, toilets, drains and effluents as a cleansing agent.

(b-4) Iodine

Chemical formula : I_2

Molecular Weight : 253.81

Standards : Iodine contains not less than 99.5% and not more than 100.5% of I_2.

Preparations

From sea weed's ash : The sea water contains iodine in the form of soluble iodates. These are absorbed by seaweeds such as *Fucus vesiculosus and Lominaria digitata.* These seaweeds are dried, burnt and obtained ash is called kelp. Iodine is extracted from water of kelp and solution is concentrated. In this process sulphate and chlorides of sodium and potassium crystallize out leaving freely soluble iodates of sodium and potassium. To this solution sulphuric acid is added and solution is decanted. The decanted mother liquid is treated with manganese dioxide and heated to collect iodine.

$$2NaI + 3H_2SO_4 + MnO_2 \longrightarrow MnSO_4 + 2NaHSO_4 + I_2 + 2H_2O$$

Physical Properties

It is a greyish violet, brittle, plates or small crystals with metallic shine, irritating odour, volatilize slowly at room temperature; soluble in glycerin, very slightly soluble in water but highly soluble in concentrated solutions of iodides.

Chemical Properties

1. Iodine forms iodates with metals like Fe, As, Sb, Mg, Pb and non metals like P, H, etc.

$$3Fe + 4I_2 \longrightarrow Fe_3I_8$$
$$2As + 3I_2 \longrightarrow 2AsI_3$$
$$Hg + I_2 \longrightarrow HgI_2$$
$$Pb + I_2 \longrightarrow PbI_2$$
$$H_2 + I_2 \longrightarrow 2HI$$
$$2P + 3I_2 \longrightarrow 2PI_3$$

2. With dilute cold alkali it forms hydroiodates while with concentrated warm alkali it forms iodates.

$$2NaOH + I_2 \xrightarrow[\text{hot}]{\text{cold}} NaI + NaIO + H_2O$$

$$6NaOH + 3I_2 \longrightarrow 5NaI + NaIO_3 + 3H_2O$$

3. Oxidation : It oxidises H_2S and $Na_2S_2O_3$

$$H_2S + I_2 \longrightarrow 2HI + S$$
$$2Na_2S_2O_3 + I_2 \longrightarrow Na_2S_4O_6 + 2NaI$$

4. It react with concentrated ammonia

$$2\,NH_3 + 3I_2 \longrightarrow NI_3 \bullet NH_3 + 3\,HI$$

5. It adsorbs on the surface of starch and starch solution turns violet.

Purity test

1. *For bromides and chlorides :* Triturate 3 gm iodine with 20 ml of water, filter, wash the filtrate, and dilute filtrate to 30 ml and add 1 gm of zinc powder. The solution decolourises filter and wash the filtrate with sufficient water to produce 40 ml to 10 ml of this solution, add 3 ml of 10 M ammonia and 6 ml of silver nitrate solution, and dilute to 20 ml with water. To 10 ml of filtrate add 1.5 ml of nitric-acid. The opalescence produced should not be more than opalescence produced by mixing 0.25 ml of 0.01 M hydro chloric acid (not more than 250 PPM).

Incompatibility

It is incompatible with alkalies and alkali carbonates.

Storage : It should be stored in earthenware containers having waxed bungs or in ground glass stoppered bottles.

Assay : It is assayed by direct iodimetric type of redox titration. Iodine is weighed in iodine flask containing 1 gm potassium iodide of 2 M acetic acid and 50 ml of water. Titrate with 0.1 N sodium thiosulphate using starch as indicator.

Factor : Each ml of 0.1 M $Na_2S_2O_3 \equiv$ 0.01263 gm of Iodine.

$$2Na_2S_2O_3 + I_2 \longrightarrow Na_2S_4O_6 + 2NaI$$

Uses : Iodine is used as an antimicrobial agent acts by formation of hypoiodous acid (HIO) which is six times more effective than hypochlorous acid (HOCl). It is used for disinfecting unbroken skin, application to mucous membrane in pharyngitis and follicular tonsillitis in combination with glycerol.

Iodine deficiency causes goitre due to iodine deficiency in human diet. For this purpose iodized salts with sodium iodides are given but should be avoided in pregnancy and lactation.

(b-5) Iodine Solution

There are various types of Iodine solutions used as germicidal action.

1. Aqueous Iodine solution B.P./I.P.(1966) (strong iodine U.S.P)
2. Weak Iodine solution B.P./I.P. (1966).
3. Strong Iodine solution I.P. (1966)
4. Iodine Tincture U.S.P.

1. *Aqueous Iodine solution B.P./I.P.(1966) (strong-Iodine U.S.P)* : It is also called as Lugol's solution. It does not contain alcohol.

Ingredients

Iodine	50 gm or 5.0% w/v
Potasium iodide	100 gm or 10.0% w/v
Purified water (q.s)	1000 ml or 100% w/v

Uses : It is a good source of iodine. It is also used as a germicide and fungicide.

2. *Weak Iodine solution*

Synonym : Iodine Tincture.

Ingredients

Iodine	20 gm or 2.0% w/v
Potassium iodide	25 gm or 2.5% w/v
Alcohol (q.s.)	1000 gm or 100% w/v

Uses : It is used as a antiseptic.

3. *Strong-Iodine solution I.P. (1966)*

Ingredients

Iodine	100 gm or 10.0% w/v
Potassium Iodide	60 gm or 6.0% w/v
Purified water	100 ml or 10.0% w/v
Alcohol (90%) q.s. to	1000 ml or 100.0% w/v

Uses : It is used as a antiseptic

4. *Iodine Tincture U.S.P.*

Ingredients

Iodine	20 gm or 2.0% w/v
Sodium iodide	24 gm or 2.4% w/v
Purified water	100 ml or 10.0% w/v
Alcohol (90%) q.s. to	1000 ml or 100.0% w/v

Iodine solutions are prepared by triturating iodine with potassium iodide in a minimum quantity of water. In this process higher iodates are formed. These higher iodates are water soluble, at this stage other ingredients are added and volume is make up with the solvent.

All iodine preparations are contra-indicated in patients exhibiting allergy or hypersensitivity to iodine.

$$KI + I_2 \longrightarrow KI_2 + I^-$$

$$KI_2 + I_2 \longrightarrow KI_3 \longrightarrow KI_4 \longrightarrow KI_5$$

(b-6) Iodine - Povidone

Standard : It contains not less than 9% and not more than 12% of available iodine calculated with reference to dried iodine.

Preparation

It is a complex of iodine with a polymer povidone [poly (2-oxo, pyrrolidin-1-ethylene)]. This complex is iodophor and acts as solubilizing agent.

Physical Properties

It is yellowish brown, amorphous powders, characteristic iodine odour; soluble in water and 95% ethanol but practically insoluble in chloroform and acetone.

Storage : It is stored in airtight iodine resistant containers.

Uses : Povidone iodine is an iodophore with a prolong anitbacterial and antifungal action. A 10% solution is used for cleansing wounds, and a 7.5% detergent solution is used for pre-operative scrubbing and antiseptic for surgeons hands. A 5% solution is used as aerosol spray over wounds, 1% solution is used for mouth wash and gargles for the treatment of mouth and throat infection. The solution of povidone iodine is also used for vaginal candidiasis.

(b-7) Boric Acid

Chemical formula : (H_3BO_3)

Molecule Weight : 61.83

Synonym : Boracic acid.

Standards : It contains not less than 99.5% and not more than 100.5% of boric acid with reference to dried substance.

Preparations

1. Industrial Process : It is prepared from the ore colemanite ($Ca_2B_6O_{11}.5H_2O$). The ore is finely powdered; dissloved in boiling water and sulphur dioxide gas is passed in the solution. The calcium bisulphite is soluble in water and remains in the solution while boric acid crystallizes out.

$$Ca_2B_6O_{11} + 4\,SO_2 + 11\,H_2O \longrightarrow 2\,Ca(HSO_3)_2 + 6\,H_3BO_3$$

(Calcium-bi-sulphite) + (Boric acid)

2. Laboratory Process : Boric acid is prepared by boiling solution of borax (37%) with 20% sulphuric acid. The solution is filtered and crystallized. The crystals of boric acid are separated and dried at room temperature.

$$Na_2B_4O_7 + H_2SO_4 + 5\,H_2O \longrightarrow Na_2SO_4 + 4\,H_3SO_4$$

Physical Properties

It is a white crystalline colourless shiny plate, uncutaneous to the touch or white crystals, odourless; soluble in boiling water and in 95% ethanol; freely soluble in boiling water, boiling 95% alcohol and glycerin.

Chemical Properties

1 (a) Heating : On heating it looses water and is converted into meta boric acid.

$$H_3BO_3 \xrightarrow{100^oC} HBO_2 + H_2O$$

(b) When it is heated at 160ºC it forms pyroboric acid and finally borone trioxide, which has glassy appearance.

$$4\ HBO_2 \xrightarrow{106^oC} H_2B_4O_7 + H_2O$$

(Pyro boric acid or tetra boric acid)

$$H_2B_4O_7 \xrightarrow[106^oC]{\text{more then}} 2\ B_2O_3 + H_2O$$

(Borone trioxide)

2. With alkali it forms salt

$$NaOH + H_3BO_3 \longrightarrow NaBO_2 + 2\ H_2O$$

(Sod. meta borate)

$$2\ NaOH + 4\ H_3BO_3 \longrightarrow Na_2B_4O_7 + 7\ H_2O$$

Borax

3. Action with calcium flouride and conc. sulphuric acid can forms boron trifluoride.

$$2\ CaF_2 + 3\ H_2SO_4 + 2\ H_3BO_3 \longrightarrow 3\ CaSO_4 + 2BF_3 + 6\ H_2O$$

4. Action with soda ash liberate carbon dioxide and borax is formed.

$$4\ H_3BO_3 + Na_2CO_3 \longrightarrow Na_2B_4O_7 + CO_2 + 6\ H_2O$$

Borax

5. It reacts with alkali acid fluorides to form sodium fluroborate.

$$2\ NaHF_2 + H_3BO_3 \longrightarrow NaBF_4 + NaOH + 2\ H_2O$$

(Sod. fluroborate).

6. With alcohol : Ethanolic solution of boric acid in ethanol is heated to give ethyl borate. This ethyl borate burns with characteristic green edged flame.

$$H_3BO_3 + 3\ C_2H_5OH \longrightarrow B\ (C_2H_5O)_3 + 3\ H_2O$$

ethyl borate.

Storage : It should be stored in well-closed containers.

Assay : It is a very weak acid having pK_a (-9.19) for the ionisation of the first proton at 25°C. This lower pK_a will not allow to carry out acid base titration. To increase the pK_a value the boric acid is esterified with polyhydric compounds such as glycerol, mannitol and catechol. This produces hydronium ion (H_3O^+), which helps in increase of Pk_a value and acid-base titration can be performed with esterified product.

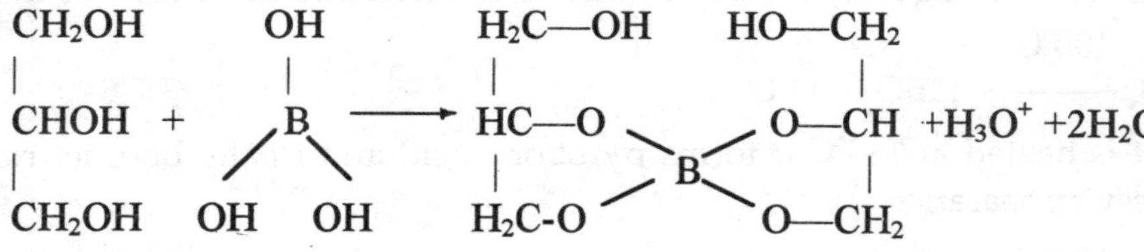

It is assayed by direct acid base type of titration. Weigh accurately about 2gm boric acid and dissolve in 150 ml of 66.6% v/v glycerin solution in water, previously neutralised with phenolphtalein. Titrate the above solution with 1 N NaOH using phenolphtalein as indicator. The standard NaOH is standardized with primary standard anhydrous succinic acid.

Factor : Each ml of 1 N NaOH ≡ O. 06183 gm of boric acid.

Uses : The aqueous solutions have been used as mouth washes, gargles, eye lotion, skin lotion, douches for irrigating the bladder and vagina, hot fomentations for whitlows, boils, carbuncles and buffer solution for external use.It is dangerous if ingested, hence should not be used for internal preparations.

(b-8) Borax

Chemcial formula : $Na_2B_4O_7 \cdot 10\ H_2O$

Molecular Weight : 381.4

Synonyms : Sodium tetraborate, Disodium tetraborate, Sodium pyroborate.

Preparations : Borax is obtained by kunite from lakes of Tibbet and India by simple crystallization of Kunite water.

Physical Properties

It is a transparent, colourless or white crystalline powder; odourless; saline or alkaline taste; efflorescent in dry air; looses water of crystallization on iginition; soluble in water and glycerol and practically insoluble in alcohol.

Chemical Properties

1. *On hydrolysis* : It produces sodium metaborate which on further hydrolysis gives boric acid and sodium hydroxide.

$$Na_2B_4O_7 + 3H_2O \longrightarrow 2\ NaBO_2 + 2H_3BO_3$$

Sodium metaborate

$NaBO_2 + 2H_2O \longrightarrow NaOH + H_3BO_3$

Boric acid

$Na_2B_4O_7 + 7H_2O \longleftrightarrow 4H_3BO_3 + 2NaOH$

2. *Heating* : On heating it losses water of crystallization above its M.P. and swells up to form porous mass, which on further heating melts and solidifies to a transparent glossy mass.

$Na_2B_4O_7 \bullet 10H_2O \longrightarrow Na_2B_4O_7 + H_2O$

$Na_2B_4O_7 \longrightarrow 2\,NaBO_2 + B_2O_3$

Sodium meta borate — Boric anhydride

|—— glassy mass——|

3. It reacts with conc. HCl or conc. H_2SO_4 to form boric acid

$Na_2B_4O_7 + 2HCl + 5H_2O \longrightarrow 2\,NaCl + 4H_3BO_3$

$Na_2B_4O_7 + H_2SO_4 + 5H_2O \longrightarrow Na_2SO_4 + 4H_3BO_3$

4. On heating with ammonium chloride if form boron nitrite.

$Na_2B_4O_7 + 2\,NH_4Cl \longrightarrow 2NaCl + 2BN + B_2O_3 + 4H_2O$

Incompatibility : Borax is incompatible with mercuric chloride, zinc sulphate, other metallic salt and also precipitate alkaloids from their solutions. e.g. : Cocaine HCl solution is precipitated by borax.

Storage : It should be stored in air-tight containers in a cool place.

Uses : It is a weak antibacterial agent simillar to boric acid. It is used in the treatment of inflammatory conditions of eye, gargles and mouth washes, lozenges for astringent action (ingredient in compound thymol glycerin lozenges).

(b-9) Silver Nitrate

Chemical formula : $AgNO_3$.

Molecular Weight : 169.89

Standards : It contains not less than 99% and not more than 100.5% of silver nitrate.

Preparations : Silver nitrate is prepared by the action of hot concentrated nitric acid on silver metal. The solution is heated to dryness till silver nitrate crystallizes.

$Ag + 2HNO_3 \longrightarrow AgNO_3 + NO_2 + H_2O$

Silver metal — Silver nitrate

Physical Properties : It is a white crystalline powder or colourless crystals; odourless; very soluble in water, soluble in 95% ethanol but very slightly soluble in ether and glycerol.

Chemical Properties

1. On heating above its melting point, oxygen is liberated.

   $2AgNO_3 \longrightarrow 2AgNO_2 + O_2$

2. When salt is heated until redness, silver metal separate.

   $2AgNO_3 \longrightarrow 2Ag + 2NO_2 + O_2$

3. On reaction with ammonia it gives brown PPt. of silver oxide.

   $2AgNO_3 + 2NH_4OH \longrightarrow Ag_2O + 2NH_4NO_3 + H_2O$

4. With ammonium thiocynate it forms reddish yellow precipitate of silver thiocynate.

   $AgNO_3 + NH_4SCN \longrightarrow \underset{\text{Reddish yellow}}{AgSCN} + NH_4NO_3$

Purity test : *For Aluminium, bismuth, copper and lead* : Dissolve 1 gm silver nitrate in a mixture of 4 ml strong ammonia solution and 6 ml of water. The resulting solution should be clear.

Incompatibility : It is incompatible with alkalies, halogen acid and their salts, phosphates and tannins.

Storage : It should be stored in light resistant, non-metallic containers. Its solution should be freshly prepared and supplied in amber coloured bottles.

Assay : It is assayed by direct argenometric or precipitation type of titration. Weigh accurately about 0.3 gm of silver nitrate and dissolved in water, add 2 ml of dilute nitric acid and 2 ml of ferric ammonium sulphate solution as indicator, and titrate with standard 0.1 M ammonium thiocyanate till reddish yellow colour is produced.

Factor : Each ml of 0.1 M Ammonium thiocyanate ≡ 0.01699 gm of silver nitrate.

$$\underset{\text{Silver nitrate}}{AgNO_3} + NH_4SCN \longrightarrow \underset{\text{Silver thiocyanate (reddish-yellow)}}{AgSCN} + NH_4NO_3$$

Uses : Silver nitrate is a caustic, astringent and bacteriocidal. It acts by precipitation of protein. The caustic activity is used to destroy warts and other small skin growths. The antibacterial activity is found in 0.1% to 1.0% solution. These are used to apply on burns (to protect from other infection), in eye (preventing opthalmic-neonatrum), irrigation fluid (urethra and bladder) etc. The prolonged use of silver preparations may results in darkening of skin due to deposition of free silver under epidermis (Argyria).

(b-10) Mild Silver Protein

Standards : It contains not less than 19% and not more than 23.0% of silver.

Physical Properties : It occur as brown powder; hygroscopic; odourless; slightly soluble in water and forms dark brown solution, insoluble in alcohol, ether and chloroform.

Storage : The solution of mild silver protein should be stored with suitable stabilizer (e.g. calcium edetate) which act as chelating agent for release of silver ions. The powder is stored in well closed container protected from moisture and light.

Uses : It is a colloidal silver in combination with proteins. It has similar use to that of silver nitrate but it is more useful because of non irritant and antibacterial on mucous membrane. It is also used in prophylaxis and treatment of respiratory tract infections.

(b-11) Mercury

Chemical formula : Hg

Molecular Weight : 200.6

Synonyms : Hydragyrum, Quick-silver.

Standards : It contains not less than 99.5% of mercury.

Preparations : It is prepared from ore cinnabar (HgS). The ore is roasted with air to remove sulphidal group as sulphur dioxide.

$$HgS + O_2 \longrightarrow \underset{\text{Mercury}}{Hg} + SO_2$$

The liberated metal is purified by dropping mercury into a column of dilute nitric acid to remove basic impurities.

Physical Properties : It is a shiny, silvery white, heavy liquid, easily broken into globules and mobile; readily volatilised on heating; practically insoluble in water, alcohol, hydrochloric acid; readily soluble in nitric and sulphuric acid.

Chemical Properties

1. When mercury reacts with nitric acid it produces mercuric salts.

$$3Hg + 8HNO_3 \longrightarrow \underset{\text{Mercuric nitrate. (mercuric-salts).}}{3Hg(NO_3)_2} + 2NO + 4H_2O$$

2. When metallic mercury is reacted with mercuric nitrate, it demonstrates auto oxidation reduction and forms mercurous nitrate.

$$Hg(NO_3)_2 + Hg \longrightarrow \underset{\text{Mercurous nitrate}}{Hg_2(NO_3)_2}$$

3. When mercury is treated with hot conc. sulphuric acid, it forms mercuric sulphate and sulphur dioxide is evolved.

$$\underset{\text{Mercury}}{Hg} + 2H_2SO_4 \longrightarrow \underset{\text{Mercuric-sulphate}}{HgSO_4} + SO_2 + H_2O$$

Uses : Metallic mercury was used as cathartic and parasiticides but it is extremely poisonous and even very minute inhalation of its vapour may lead to death. Various derivatives of mercury are used as germicidal, antibacterial, anti infectious and diuretic.

(b-12) Yellow Mercuric Oxide

Chemical Formula : HgO

Molecular Weight : 216.59

Synonym : Yellow precipitate.

Standards : It contains not less than 99.5% of mercuric oxide with reference to dried HgO at 105°C for 1 hour.

Preparations : It is prepared by pouring mercuric chloride solutions into sodium hydroxide solution slowly. The mixture is kept aside for 1 hour for complete precipitation and for uniform orange yellow colour.

$$HgCl_2 + 2NaOH \longrightarrow Hg(OH)_2 + 2\,NaCl$$

$$Hg(OH)_2 \longrightarrow \underset{\text{Yellow mercuric oxide}}{HgO} + H_2O$$

The supernatant liquid is poured off and the precipitate is washed with water till it becomes free from alkali. The precipitate is dried in dark place at a temperature not exceeding 30°C.

Physical Properties : It is a orange yellow, heavy, amorphous powder; odourless; practically insoluble in water and alcohol but readily soluble in hydrochloric and nitric acid.

Storage : It should be stored in well-closed containers protected from light and moisture. It slowly gets converted to red mercuric oxide (red-oxide) on exposure to light.

Uses : It is a local anti-infective, anti-bacterial used in mild infections, inflammations and conjunctivitis of eyes. It is also used in the treatment of fungal infections of skin and infestation of body lice.

(b-13) Ammoniated Mercury

Chemical Formula : $NH_2\,Hg\,Cl$

Molecular Weight : 252.10

Synonyms : White precipitate, Aminochloride of mercury, Hydaragyrum Ammoniatum.

Standards : It contains not less than 98.0% w/w NH_2HgCl.

Preparations : It is prepared by the treatment of 5.0% mercuric chloride solution with 20.0% dilute ammonia solution. The precipitate is collected, washed with cold water and dried below 30°C.

$$HgCl_2 + 2\,NH_3 \longrightarrow \underset{\text{(PPt.)}}{NH_2HgCl} + NH_4Cl$$

Physical Properties : It is a white powder; odourless; practically insoluble in water, alcohol and ether; soluble in warm hydrochloric, nitric, and acetic acids.

Storage : It should be protected from light.

Uses : It is a mild antiseptic. It is applied to the skin surrounding the perineum skin to destroy threadworms and to reduce reinfection. It is also used in the treatment of low grade staphylococcal infections of the skin and in psoriasis. It is also used in the treatment of impetigo contagiosa and treatment of crab louse infestations. Excess use of ammoniated mercury develops chronic toxicities therefore prolonged use is not recommended. 5% ointment and 3% opthalmic ointment are available for antimicrobial action.

(D-c) SULPHUR AND ITS COMPOUNDS

Sulphur is an important component in various synthetic and natural medicinal compounds. It is distributed in volcanic area, especially in Italy and United States. It is found in a free state or as a complex with various metals like Iron pyrite (FeS_2), Gypsum, ($CaSO_4 \bullet 2H_2O$) and Baryte, ($BaSO_4$). It is also found with valuable metals like antimony, mercury, bismuth, lead, zinc and molybdenum. These ores are converted into hydrogen sulphide with the help of inorganic acids and hydrogen sulphide is oxidised to give sulphur.

Sulphur exist in various allotropic forms like solid as Rhombic (α-Sulphur, α S, Sublimed-sulphur or Flower of sulphur), and Monoclinic (β-sulphur, β S); liquid as λ sulphur (S λ), μ -sulphur (Sμ); plastic sulphur (formed on super cooling μ-Sulphur) and amorphous sulphur (formed on aging of "Plastic" sulphur).

Sulphur is used as cathartic, antiscabetic and antibacterial agent. It acts by oxidation reduction products of sulphur like sulphur dioxide, sulphites, pentathonic acid. Sulphur is also used as scabicide. It also acts as a depilator or in softenning of hair by combining with disulphide linkages in amino acids present in hair.

(D-c.1) PRECIPITATED SULPHUR

Chemical Formula : S

Molecular Weight : 32.60

Synonyms : Milk of sulphur, Precip. sulphur.

Preparations : When sodium thiosulphate is acidified with dilute mineral acids like hydrochloric acid, it forms thio sulphuric acid ($H_2S_2O_3$) which rapidly decomposes to sulphur dioxide and precipitates sulphur.

$$Na_2S_2O_3 + 2HCl \longrightarrow \underset{\text{Thio sulphuric acid}}{H_2S_2O_3} + 2NaCl$$

$$H_2S_2O_3 \longrightarrow \underset{\text{Precipitated sulphur}}{S} + SO_2 + H_2O$$

Physical Properties : It is pale greyish yellow or greenish yellow; odourless soft powder free from grittiness, practically insoluble in water and alcohol; completely soluble in carbon disulphide; soluble in hot aqueous alkali hydroxide solutions.

Uses : It is a mild antiseptic,parasiticides used in the treatment of acne and scabies. Sulphur ointment is used as a superficial fungicidal.

It is a mild laxative. It gets converted into alkali sulphides and causes irritant action in small intestine. Sulphur is used in the treatment of abnormal secretions of sebum from sebaceous glands producing oily or scaly skin. It also loosens or softens horny, scaly sites from the skin.

(D-c.2) SUBLIMED SULPHUR

Chemical Formula : "S".

Molecular Weight : 32.06

Standard : Flower of sulphur, sub. sulphur.

Preparations : When sulphur is heated and vapours are cooled at suitable temperature, it forms crystalline friable mass. This yellow powder is sublimed sulphur. This is collected and sieved.

Physical Properties : It is a yellow, fine gritty powder; faint odour; practically insoluble in water and alcohol but partially soluble in CS_2 .

Storage : It should be stored in well-closed containers.

Uses : It has similar therapeutic uses as that of precipitated sulphur.

Dose : 1 to 4 gms

(D-c.3) SELENIUM SULPHIDE

Chemical formula : SeS_2

Molecular Weight : 143.09

Synonym : Selenium disulphide.

Preparation : It is prepared by the treatment of selenious acid with saturated solution of hydrogen sulphide in presence of aluminium chloride. Aluminium chloride acts as a coagulant which precipitate selenium sulphide and is filtered; washed and dried.

$$\underset{\text{Selenious acid}}{H_2SeO_3} + 2H_2S \longrightarrow \underset{\text{Selenium sulphide}}{SeS_2} + 3H_2O$$

Physical Properties : It is a bright orange to reddish brown powder; fowl odour; practically insoluble in water, alcohol and organic solvent.

Storage : It should be stored in well-closed containers.

Uses : It is highly toxic if taken internally. But it absorbs in traces, when applied to the skin as cream, suspensions to control dandruff and seborrhoeic dermatitis of the scalp. It is applied 5- 10 ml over the scalp with warm water. It is applied twice a week, later once a week and then as per the condition of disease. The hairs should be thoroughly rinsed and all the traces of suspension are removed with a finger nails of the hand to prevent it from eye or mouth. The excessive use of selenium sulphide may cause loss of hairs.

(D-d) ASTRINGENT

Astringents are applied topically and precipitate the protein and reduce the cell permeability. They mainly constricts smooth muscles e.g., blood capillaries etc. These agents have limited power of penetration and lacking deeper effects, Hence they are safely used on human.

Astringents are used as (i) To prevent diarrhoea caused by surface infection on inner walls of intestine; (ii) Styptic : Promoting coagulation and constricting of small capillaries, and prevent bleeding from cuts; (iii) It constrict pores on the skin to decrease sweating (antiperspirent); (iv) Promote healing and harden the skin; (v) Restrict blood supply to inflamed part; (vi) As a deodorant by acting with lipids of sweat and reduces bacterial growth.

(D-d.1) ALUM

Chemical formula : $AlK(SO_4)_2 \cdot 12H_2O$

Molecular weight : 474.40

Synonyms : Potash alum, potassium aluminium sulphate.

Preparations : Alum is prepared by addition of hot concentrated solution of potassium sulphate to a hot solution of aluminium sulphate. This solution is concentrated and cooled to crystallize out alum.

$$Al_2(SO_4)_3 + K_2SO_4 + 2H_2O \longrightarrow \underset{\text{Alum.}}{2KAl(SO_4)_2} + 12H_2O$$

Physical Properties : It is an odourless, colourless or transparent crystalline granular powder; sweetish astringent taste; soluble in water and glycerol but practically insoluble in alcohol.

Chemical Properties

1. Solution of alum in water is acidic and 10% solution shows pH between 3-3.5.
2. When it is heated, it looses water of crystallization and forms anhydrous alum. It melts at 200°C.

$$KAl(SO_4)_2 \cdot 12H_2O \xrightarrow{\text{Heating}} \underset{\text{Anhydrous alum}}{KAl(SO_4)_2} + H_2O$$

Incompatibility : Alum is incompatible with borax, alkali hydroxides, carbonates, phosphates and salts of calcuim, lead, mercury and tannin.

Storage : It should be stored in airtight containers.

Uses : It is a powerful astringent. This property is used as mouthwashes and gargles in 1-4% dilute solution but have destructive action on teeth. A 5% alum solutions are used by athletes to harden skin of feet and 15% powder in talc is used as foot powder. It is used as a styptic for minor cuts and abrasions. It is used as protein precipitant in the preparation of precipitated diphtheria and tetanus toxoids.

(D-d.2) ZINC SULPHATE

Chemical formula : $ZnSO_4 \bullet 7H_2O$

Molecular Weight : 287.54

Preparation : It is prepared by the heating of zinc metal with dilute sulphuric acid, till liberation of hydrogen gas stops. Then solution is filtered. Filtrate is concentrated to crystallize zinc sulphate.

$$Zn + H_2SO_4 \xrightarrow{\text{Heating}} ZnSO_4 + H_2$$

Physical Properties : It is a colourless, transparent or white crystals; odourless; very soluble in water, practically insoluble in 95% ethanol; and efflorescent in nature.

Chemical Properties

1. Heating : Water of crystallization is lost with an increase in temperature as follows.

$$ZnSO_4 \bullet 7H_2O \xrightarrow{70^0c} ZnSO_4 \bullet 6H_2O + H_2O$$

$$ZnSO_4 \bullet 7H_2O \xrightarrow{100^0C} ZnSO_4 \bullet H_2O + 6H_2O$$

$$ZnSO_4 \bullet 7H_2O \xrightarrow{200^0C} ZnSO_4 + 7H_2O$$

$$ZnSO_4 \bullet 7H_2O \xrightarrow{800^0C} ZnO + SO_2 + O_2$$

Incompatibility : It is incompatible with alkali carbonates and hydroxides and with astringent infusions and decoctions.

Storage : It should be stored in tightly closed, non-metallic containers.

Uses : Zinc sulphate is used as astringent in ophthalmic preparation (0.25% aqueous solution). Its powder and lotions also shows antiseptic and astringent property.

It is used as reflex emetic when used internally. It is recommended in narcotic poisoning.

(E) DENTAL PRODUCTS

The poor dental health is a very common problem in third world countries. It is due to ignorance, illiteracy and habits of chewing betel leaves, tobacco and pan masalas (Gutaka). The tooth consists of three layers of calcified tissues.

i. **Enamel** : It is white, outer most, hard covering present on the tooth projecting above the gum. It contains 98% of mineral and is very dense.

ii. **Cementum** : It is the portion of tooth which is fixed in the gums and are not seen from outside. It is a bone like structure.

iii. **Dentine** : It is the main pulp cavity present in entire tooth. The external shape of dentine, differentiates the tooth. It consists of 75% of mineral. It is hard and dense.

The common problems associated with tooth are the formation of cavities, cleaning of tooth and reduction of shinning over tooth. These problems are over come by the use of Anti-caries agents, Dentrifrices and Polishing agents.

Anticaries agents

The probable cause of caries is associated with the diet of the people consuming high fermentable carbohydrates. There is no definite co-relation between inorganic calcium level and dental caries. Caries occur due to action of lactic acid obtained from bacterial metabolism of dietary carbohydrate.

The build-up of plaque (place over tooth where bacteria can grow) over the tooth surface helps in decay process by forming cavities on the tooth surface, in which food particles can be degraded by the bacteria present in the mouth. The brushing of tooth prevents lodging of food particles over tooth surface and helps in hygenic teeth. The ammoniated tooth pastes are supposed to neutralize lactic acid formed by *lacto bacillus* bacteria and also decreases bacterial count. Some antibiotics are also tried as anticaries agent but the problem of sensitization and bacterial resistance to antibiotic are some of the problems which restricts its application.

Fluorides are brushed with other agents which helps in preventing caries. The use of fluoride solutions as internal solutions are well absorbed in gastrointestinal tract and also effective as anticaries, but fluorides are deposited in bone or in developing teeth causes dental fluorosis (Mottled-enamel) in excessive fluoride ingestion. These fluorides acts by either bacterial inhibition or by decreasing acid solubility of enamel which can be achieved by either reduction in number of defects in apatite-crystals or by combining with carbonates during apatite formation.

Only one PPM fluoride in drinking water is recommended for anti-caries activity, which can be reduced in summer months at tropical regions (hot places) because of more consumption of water. Sodium fluoride and stannous fluoride are incorporated in pure drinking water to achieve fluoride level of one PPM.

The nonfluoridate water may cause osteodystrophy in patients with receiving long term renal dialysis. The internal intake of fluorides is considered to be better than topical application as anticaries. Absorbed fluoride comes out with saliva and acts on tooth as anticaries.

(E-1) SODIUM FLUORIDE

Chemical formula : NaF

Molecular Weight : 41.99.

Standards : It contains not less than 98.5% and not more than 100.5% of NaF with reference to dried substance.

Preparations

1. It is prepared by treatment of hydrofluoric acid with sodium carbonate in lead container.

$$2HF + Na_2CO_3 \longrightarrow 2NaF + H_2O + CO_2$$

2. It is also prepared by the double decomposition of calcium fluoride with sodium carbonate, the insoluble calcium carbonate is filtered out.

$$CaF_2 + Na_2CO_3 \longrightarrow 2NaF + CaCO_3$$

Calcium fluoride

Physical Properties : It occurs as a colourless crystals or white powder; odourless; soluble in water and insoluble in ethanol (95%).

Purity test : Test for acidity or alkalinity : The boiling solution of sodium fluoride is titrated with 0.1 M sodium hydroxide until a red colour is produced. Not more than 1.5 ml of 0.1 M NaOH is required.

Storage : It should be stored in air tight containers.

Uses : Sodium fluoride is used in the treatment of prophylaxis of dental caries in communities where the intake of fluoride from drinking water and food is minimum. Sodium fluoride in 2% aqueous solution is used topically or applied to children's teeth. Sodium fluoride is used as a constituent of some insecticides and rhodenticides.

(E-2) STANNOUS FLUORIDE

Chemical formula : SnF_2

Molecular Weight : 156.69

Synonym : Tin Fluoride

Preparations

1. From stannous oxide : It is prepared by evaporating a solution of stannous oxide with hydrofluoric acid in the absence of oxygen in a lead retard.

$$\underset{\text{Stannous oxide}}{SnO + 2HF} \longrightarrow \underset{\text{Stannous fluoride}}{SnF_2 + H_2O}$$

2. Tin fluoride solution is prepared by dissolving a tin fluoride capsule in water. A fresh solution (about 8%) is produced and used in dentistry.

Physical Properties

It occurs as a colourless or white crystalline powder; unpleasant or astringent salty or bitter taste; freely soluble in water, insoluble in alcohol, ether and chloroform.

Chemical Properties

1. *On hydrolysis* : When stannous fluoride is hydrolysed and allowed to stand, it slowly develop a white precipitate of stannous hydroxide.

$$SnF_2 + 2H_2O \longrightarrow \underset{\text{Stannous hydroxide}}{Sn(OH)_2} + 2\,HF$$

Aqueous solution of stannous fluoride is acidic (approx. pH 3.0).

2. *On exposure to air* : When stannous fluoride is exposed to the air, it is oxidised to stannic fluoride or stannic oxy-fluoride ($SnOF_2$).

$$\underset{\text{Stannous-fluoride}}{SnF_2} + [O] \longrightarrow \underset{\text{Stannous oxyfluoride}}{SnOF_2}$$

$$\underset{\text{Stannous fluoride}}{SnF_2} + [O] \longrightarrow \underset{\text{Stannic fluoride}}{SnF_4} + \underset{\text{Stannous oxide}}{SnO}$$

Uses : Stannous fluoride is used topically . It is used for prevention of caries and seems to be superior to sodium fluoride because it has a simplified application and greater effectiveness. A freshly 8% aqueous solution of stannous fluoride is applied to the cleaned dry teeth every 6 to 12 months, where as a 2% freshly aqueous solution of stannous fluoride is applied to the clean, dry teeth four times during a period of 10 days.

Dentrifrices

Dentrifrices are also called cleaning agents. These agents are generally coarse to fine powder with abbrassive property. The abrasiveness should be evaluated with respect to removal of stain from tooth but should not produce any harm to the enamel layer of tooth. These agents are mixed with detergents or soap or surfactants and flavouring agents. These are non-carbohydrate substances like calcium carbonate, dicalcium hydrogen phosphate, sodium metaphosphates etc. These agents are used as ointments or powders which produces frothing, sweet taste and pleasant odour hence thickening agents like polyhydroxy alcohols (glycol, glycerol, sorbitol etc.), acacia, tragacanth, starch and cellulose derivatives are added.

The dentrifrices are mixed with desensitizer and with polishing agents. The desensitizer used are strontium chloride and zinc chloride. They decreases hyper-sensitivity of teeth especially where erosion has occured near the gum line. They also reduces sensitivity of teeth from cold and hot conditions. The polishing agent used is calcium carbonate which is very fine and gives a shining to teeth.

(E-3) CALCIUM CARBONATE

Chemical formula : $CaCO_3$

Molecular Weight : 100.09

Synonym : Calc carb, Calcii-carbons, Precipitated calcium-carbonate, Precipitated chalk.

Standards : It contains not less than 98.0% and not more than 100.5% calcium carbonate calculated as dried substances.

Preparations

1. When boiling solution of sodium carbonate is added with calcium chloride then a precipitate of calcium carbonate is formed. The precipitate is washed, filtered till filtrate become free from chloride ions, precipitate is dried below 105°C.

$$Na_2CO_3 + CaCl_2 \longrightarrow NaCl + \underset{\text{(Calcium carbonate)}}{CaCO_3}$$

2. When carbon dioxide gas passes through lime water, a milky white precipitate of calcium carbonate is formed.

$$CO_2 + Ca(OH)_2 \longrightarrow \underset{\text{Calcium carbonate}}{CaCO_3} + H_2O$$

Physical Properties : It is a white, odourless powder; very slightly soluble in water but solubility is increased by the presence of carbon dioxide or any ammonical salt soluble in dilute acids.

Chemical Properties

1. When excess of CO_2 is passed in the solution of calcium carbonate, soluble calcium bicarbonate is formed.

$$\underset{\substack{\text{Calcium carbonate}\\ \text{(Insoluble)}}}{CaCO_3 + CO_2 + H_2O} \longrightarrow \underset{\substack{\text{Calcium bicarbonate}\\ \text{(soluble)}}}{Ca(HCO_3)_2}$$

2. When it is treated with dilute acids, effervescence of carbon dioxide liberates.

$$CaCO_3 + 2HCl \longrightarrow CaCl_2 + H_2O + CO_2$$

Purity test

1. *For Barium* : To 2.6 gm of substance, add 10 ml of dilute acetic acid, boil, cool and add 10 ml of calcium sulphate solution. The solution should remain clear.

2. *For Magnesium and alkali metals* : Dissolve 1 gm substance in 10 ml of dilute hydrochloric acid and neutralize by addition of ammonia. The solution is heated to form solution of ammonium oxalates. To this add 1.5 ml of dilute sulphuric acid and evaporate to dryness on water bath, heat the residue and weight should not be more than 10 mg (1%).

Storage : It should be stored in well closed containers.

Uses : It is a very good dentrifrices because of mild abrassive quality. It is also used as antacid in various preparations.

(E-4) SODIUM METAPHOSPHATE

Chemical formula : $NaPO_3$

Molecular Weight : 102.00

Synonym : Madrell's salt.

Preparations : It is prepared by heating two parts of sodium nitrate with one part of phosphoric acid at 350°C

$$2NaNO_3 + H_3PO_4 \longrightarrow NaNH_4 \bullet HPO_4$$

$$NaNH_4.HPO_4 \longrightarrow NaH_2PO_4 + NH_3$$

$$NaH_2PO_4 \longrightarrow \underset{\text{Sod. metaphosphate}}{NaPO_3} + H_2O$$

Physical Properties : It is a white, odourless powder, practically insoluble in water but soluble in mineral salts.

Storage : It should be stored in well closed containers.

Uses : It is a good abrassive, cleaning, polishing agent and detergent.

(E-5) DIBASIC CALCIUM PHOSPHATE

Chemical formula : $CaHPO_4 \bullet 2H_2O$ ¦ $CaHPO_4$

Molecular Weight : 172.1 ¦ 136.06

Synonyms : Calci Hydrogen Phosphas, Calcium hydrogen phosphate, Dicalcium poshphate.

Standards : It contains not less than 98.0% and not more than 105.0% of $CaHPO_4$ (for anhydrous), $CaHPO_4 \bullet 2H_2O$ (for dihydrate).

Preparation : It is prepared by the action of neutral calcium chloride with disodium hydrogen phosphate.

$$CaCl_2 + Na_2HPO_4 \longrightarrow \underset{\text{Calcium hydrogenphosphate}}{CaHPO_4} + 2\,NaCl$$

Physical Properties : It is a white crystalline, odourless powder; practically insoluble in water and 90% alcohol but soluble in dilute hydrochloric and nitric acids.

Chemical Porperties : When treated with hydrochloric acid it forms monobasic calcium phosphate.

$$2CaHPO_4 + 2HCl \longrightarrow Ca(H_2PO_4)_2 + CaCl_2$$

Calcium hydorgen phosphate → Monobasic calcium phosphate

Storage : It should be stored in well closed containers.

Uses : It is a principle ingredient in tooth paste as dentrifrices. It is also given as source of calcium and phosphorous in pregnancy and lactation to calcium and phosphate deficient women.

Dose : 1 to 5 gm daily as calcium supplement.

(E-6) STRONTIUM CHLORIDE

Chemical formula : $SrCl_4$ $SrCl_4 \bullet 6H_2O$

Molecular Weight : 158.52 ¦ 266.52

Physical Properties : It is a colourless, odourless, white granules. It is effloroscent in dry air and deliquescent in moist air. It is very soluble in water and alcohol.

Storage : It should be stored in well-closed containers, protected from air and moisture.

Uses : It is a desensitizing agent used in dentrifrices.

(E-7) ZINC CHLORIDE

Chemical Formula : $ZnCl_2$

Molecular Weight : 136.29

Synonym : Butter of zinc.

Standards : It contains not less than 95.0% and not more than 100.5% zinc chloride.

Preparation : It is prepared by heating metallic zinc with concentrated HCl, until evalution of hydrogen complete.

$$Zn + 2HCl \longrightarrow ZnCl_2 + H_2$$

Zinc choride

Physical Properties : It is a white crystalline powder; odourless; deliquescent; very soluble in water and freely soluble in 95% ethanol and glycerin.

Purity test

1. *For aluminium, calcium, heavy metal, iron and magnesium* : To 8 ml of solution "A", add 2 ml of strong ammonia solution and shake. If solution is coloured,add 1 ml of a 9% w/

v solution of disodium hydrogen phosphate; the resulting solution remain clear for at least 5 minutes.

2. *Solution A* : Weigh 2 gm zinc chloride, add 38 ml of freshly boiled and cooled water. Add drop by drop 2M hydrochloric acid until solution is formed. Dilute this to 40 ml with carbon dioxide free water.

Storage : It should be stored in tightly closed non-metallic containers.

Uses : It is antiseptic, astringent to skin and mucous membrane and used as mouth washes and ulcers. A 10% solution is used as dentine desensitizer for teeth.

(F) INHALANTS

Inhalants are a type of formulation which either gas or mist reaches directly to the affected area when administered through nasal or buccal cavity. The five gases used as inhalants are oxygen, carbon dioxide, nitrous oxide, nitrogen and helium. In this chapter properties, application, significance of oxygen, carbon dioxide and nitrous oxide are described.

Role of oxygen : Any living cell requires oxygen. Oxygen is necessary for normal oxidative metabolic process in cell for the production of energy. Energy is used by cells in the form of ATP, FAD, and NADH.

Transport of oxygen from the lungs to the tissues and *vice versa* is done by haemoglobin. Haemoglobin combines with O_2 to form oxygenated haemoglobin or oxyhaemoglobin. When haemoglobin looses its oxygen it is known as reduced haemoglobin. Oxygenation is a reversible process and oxygen haemoglobin linkage can be cleaved. Formation of oxyhaemoglobin and reduced haemoglobin or the loading and unloading of oxygen from the tissues depends on (a) Pressure of oxygen (b) Pressure of carbon dioxide (c) Body temperature. It is a characteristic property of haemoglobin, that it quickly looses oxygen if pressure of oxygen (Po_2) is 40-60 mm Hg. However, if pressure of oxygen is above 80 mm Hg, their occurs 100% saturation of haemoglobin with oxygen and no further loading can occur after this. This relation between saturation of haemoglobin with oxygen can be illustrated by a oxygen dissociation curve.

Role of carbon dioxide : Venous blood carries about 52 CC% of CO_2 where as, arterial blood carries about 48CC% of CO_2. Thus about 4 CC% CO_2 is released in lungs by deoxygenated blood and from the tissues 40CC% CO_2 is transferred to oxygenated blood.

Under normal conditions of temperature and pressure only 2.7 ml of CO_2 is carried in physiological solution i.e. as carbonic acid. Majority of CO_2 is carried by blood as bicarbonate.

$$H_2CO_3 \longleftrightarrow H^+ + HCO_3^-$$

$$pH = pk_a - \log \frac{[H_2CO_3]}{[HCO_3^-]}$$

When CO_2 enters the blood, chloride ion from plasma enters the red cells, while basic sodium ion is left behind. When carbon dioxide escapes from blood, chloride ions leave the cells, enter the plasma and combine again with sodium ion. This phenomenon is known as chloride shift. It is a major CO_2 transport mechanism (45.7 ml of CO_2 is being transported by this mechanism). A part of CO_2 (3.7 ml) is also transported as carbamino compound which is formed when NH_2-radical of protein (Plasma protein or globin of haemoglobin) combines with one molecule of CO_2 (as free gas). This process does not require the enzyme carbonic anhydrase.

Exchange of gases : The exchange of gases at the cellular level in lungs and tissues, namely, oxygen (O_2) and carbon dioxide (CO_2) are transported between lungs and tissues through blood. These exchange mechanisms work on the principle of biophysics that gases try to equalise the diffusion, if there is pressure gradient.

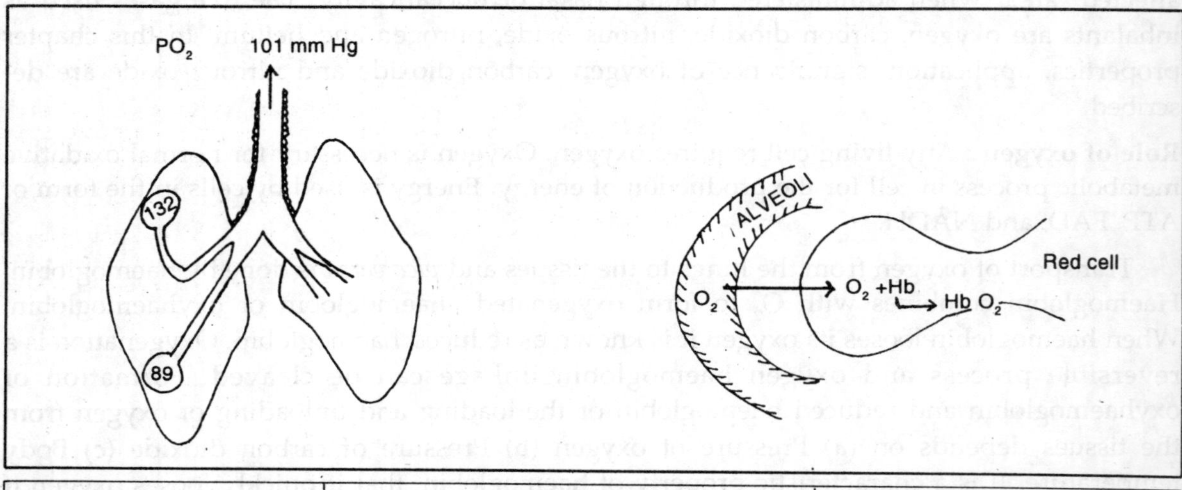

Tissues	Plasma	Erythrocytes
		CO_2 carried as
CO_2 Diffusion (Carbon di oxide)	CO_2 diffusion as	CO_2 dissolved gas H2O + Carbonic anhydrase ↓
	CO_2 carried	H_2CO_3 (Carbonic Acid) ↑
	in plasma HCO^-_3 ← as result of haemoglobin buffering	HCO^-_3 $HCO^-_3 + H^+$ CO2 carried in erythrocytes as HCO^-_3
O_2 ← (Oxygen)	O_2 ←	Fe O_2 —HbO_2 reduse to Hb→ Fe HbO_2 (Oxyhaemo-globin) Hb (Haemo globin)

Exchange of gases in alveoli or lungs : The inspired air has Po_2 (Partial pressure of O_2) in blood as 158.2 mm Hg whereas, Pco_2 (Partial pressure of CO_2) as 0.3 mm Hg. The blood coming from right ventricle of heart (i.e. through pulmonary arteries) have Po_2 as 40.0 mm Hg and Pco_2 as 46.0 mm Hg. Thus in the alveoli because of the differences in partial pressures, there is quick diffusion of O_2 from inspired air to the blood and of CO_2 from the blood to the air in the lungs. As a result the air to be expired has Po_2 as 116.0 mm Hg and Pco_2 as 26.8 mm Hg, whereas, the blood going out from lungs to heart will have high amount of O_2 (Po_2 100 mm Hg) and low amount of CO_2 (Pco_2 40 mm Hg).

Exchange of gases at tissue level : Similarly when the blood reaches various tissues there is exchange of O_2 from the blood to tissues and CO_2 from tissue to blood. The reason is the pressure gradient. The Po_2 is about 40-60 mm Hg and Pco_2 is 35-45 mm Hg at the arterial blood. At the level of lung as well as at the cellular level the gases exchange are on the basis of diffusion.

(F-1) OXYGEN

Chemical formula : O_2

Molecular Weight : 32.00

Standards : It contains not less than 99.0% v/v of oxygen.

Preparations

1. It is prepared by electrolysis of water between two iron electrodes dipped in 10% aqueous potassium hydroxide solution. The oxygen is liberated at anode and is collected.
2. The atmospheric air is purified by removing moisture, carbon dioxide dust etc. and liquid air is fractionated by distillation. Pure oxygen is filled in metallic cylinders under reduced pressure.

Physical Properties : It is colourless, odourless gas; one volume of gas is dissolved in 32 volumes of water.

Chemical Properties

1. Oxygen with non-metals forms oxides like :
2. Oxygen with metals forms oxides like :

$$4\,Fe + 3O_2 \longrightarrow 2\,Fe_2O_3$$

$$2\,Mg + O_2 \longrightarrow 2\,MgO$$

$$S + O_2 \longrightarrow SO_2$$

$$C + O_2 \longrightarrow CO_2$$

$$4\,P + 5O_2 \longrightarrow 2\,P_2O_5$$

Storage and Labelling : It should be stored in metallic cylinders with appropriate valves. Valves and tapes should not be lubricated with grease or oil. The shoulder of cylinder should be painted with white and rest with black containing a label and symbol of oxygen at shoulder.

Uses : It is essential for normal respiratory function and theraputically used in the treatment of hypoxia, artificial respiration and other serious patients.

(F-2) CARBON DIOXIDE

Chemical formula : CO_2

Molecular Weight : 44.00

Preparations

1. It is prepared by heating potassium bicarbonate at 100°C.

$$2\,KHCO_3 \xrightarrow{100^\circ C} CO_2 + K_2CO_3 + H_2O$$

Pot. bicarbonate Carbon dioxide

2. Carbon dioxide gas is prepared by the treatment of mineral acid with calcium carbonate.

$$2HCl + CaCO_3 \longrightarrow CaCl_2 + H_2O + CO_2$$

Physical Properties

1. *Liquid carbon dioxide* : It is a heavy, colourless, acidic gas, liquifies under pressure and soluble in equal parts of water at O°C.
2. *Solid carbon dioxide (dry ice)* : It is a white, solid with low thermal conductivity. It has a temperature of -80°C. On exposure it forms CO_2 (solid) snow.

Chemical Properties

1. When carbon dioxide gas is passed over lime water it forms calcium carbonate.

$$CO_2 + Ca(OH)_2 \longrightarrow CaCO_3 + H_2O$$

Calcium carbonate

2. When gas is passed in a solution of calcium carbonate, a milky white precipitate of calcium bicarbonate is formed.

$$CO_2 + Ca(HCO_3)_2 \longrightarrow CaCO_3 + H_2O$$

$$CO_2 + CaCO_3 + H_2O \longrightarrow Ca(HCO_3)_2$$

Calcium bicarbonate

Storage : Carbon dioxide should be stored in metallic cylinders at cool temperature not exceeding 31°C. The cylinder should be painted grey having a symbol of gas.

Uses Low concentration (upto 10%) is used to stimulate respiratory and cardiovascular centre but higher concentration produces convulsion, respiratory depression and may prove to be fatal.

(F-3) NITROUS OXIDE

Chemical formula : N_2O

Molecular Weight : 44.00

Synonyms : Laughing Gas, Nitrogen monoxide, Dinitrogen oxide.

Standards : It contains not less than 98% v/v of N_2O.

Preparation : It is prepared by heating ammonium nitrate. The gas is purified by washing with sodium chromate, NaOH and water.

$$NH_4NO_3 \longrightarrow \underset{\text{Nitrous oxide}}{N_2O} + 2H_2O$$

Physical Properties : It is a colourless; odourless gas; freely soluble in ethanol; soluble in ether and oil and one volume of gas dissolves in 1.5 volume of water.

Storage and labelling : It should be stored under pressure in metallic cylinders, painted blue and carries a label stating nitrous oxide.

Uses : It is a analgesic, anaesthetic, used in dental extraction, minor operations and calming excited mental patients. It is inhaled with 20-40% of oxygen.

Hazards : Nitrous oxide can be responsible for excitement of patient after anesthesia leading to delirient effects and the patient may charge the physician of a crime not committed by him. It is also slightly inflammable and explosive thereby causing accidents in the operation theatres or at any place where it is used.

(G) RESPIRATORY STIMULANTS

The respiratory stimulating agents have a reflux action on the patient taking sudden deep breath. These are used in case of unconscious or fainted person. The common respiratory stimulants used are Aromatic spirit of Ammonia (N.F.), strong ammonia, carbon dioxide and oxygen. In aromatic spirit of ammonia. The principal ingredient used is "Smelling salt"(Ammonium Carbonate).

AMMONIUM CARBONATE

Chemical Formula : $(NH_4HCO_3)m\ (NH_2CO_2NH_4)n$

Molecular Weight : Indefinite [(79)m + (78)n]

Synonyms : Ammonium sesquicarbonate Savlvolatile, Preston's salt, Hartshron.Smelling salt.

Standards : The mixture of ammonium bicarbonate and ammonium carbonate should contain not less than 30% equivalent to ammonia.

Physical properties : It is a translucent crystalline substance containing strong ammonical (Pungent) odour and taste. On exposure to air it dissociate and volatilizes and becomes porous lumps or white powder. It is freely soluble in water and partly soluble in alcohol.

Incompatibility : Ammonium carbonate is incompatible with acid salts of iron and salts of alkaline earth metals.

Storage : It should be stored in air-tight containers in a cool place.

Uses : It is irritant to respiratory mucous membrane. The official preparation (Aromatic spirit of ammonia) contain 0.25% w/v Ammonium carbonate. It is also important constituent (47% w/v) in strong ammonium acetate soln.

(H) EXPECTORANTS AND EMETICS

Expectorants : These agents enhance the secretion of sputum from the air passage so, it is easier to cough up. Expectorants act by increasing the bronchial secretion or making it less viscous. They irritate the lining of stomach which provides stimulus for the reflux production of sputum by the glands in the bronchial mucous membrane. The agents commonly used are ammonium chloride, glyceryl guaiacolate, potassium iodide, hydroiodic acid syrup and syrup of Ipecac.

Emetics : These agents produce irritation on G.I.T muscles or stimulate chemoreciptor trigger zone (C.T.Z) located in medulla oblongate and produce vomitting. Emetics are contraindicated in patients suffering from shock, C.N.S depression, heart disease, tuberculosis and advance pregnancy. Emetics are also mixed with expectorant to expel the cough.

(H-1) AMMONIUM CHLORIDE

Chemical formula : NH_4Cl

Molecular Weight : 53.49

Synonyms : Ammonium muriate, Sal ammoniac, Salmiac, Amchlor, Ammon chlor.

Standards : It contains not less than 99.5% dried ammonium chloride.

Preparation : It is prepared by neutralizing hydrochloric acid with ammonia. After neutralization the solution is evaporated to dryness, the crude product is recrystallizes or purified by sublimation.

$$NH_3 + HCl \longrightarrow NH_4\,Cl$$

Physical properties : It is a colourless or white crystalline powder, odourless and saline in taste. It is soluble in 2.7 parts of water and in 100 parts of alcohol at 20°C. Its five percent solution having a pH 4.5 to 6.

Purity tests

1. *For Barium* : Dissolve 0.5 gram ammonium chloride in 10 ml of water, add 1 ml of dil sulphuric acid, no turbidity is produce within two hours.

2. *For Thiocyanate* : Acidify 10 ml of 10% w/v soln. with hydrochloric acid and add a few drops of ferric chloride solution, no red colour is produced.

Assay : It is assayed by acid base titration. The neutral formaldehyde solution is added so that ammonium chloride will be converted into hexamine and hydrochloric acid. The liberated acid is titrated with 0.1 N NaOH using phenophthalein as indicator.

Factor : Each ml of 0.1 N NaOH ≡ 0.05349 g of Ammonium chloride.

$$NH_4Cl + HCHO \longrightarrow Hexamine + HCl$$

$$HCl + NaOH \longrightarrow NaCl + H_2O$$

Incompatibility : It is incompatible with alkali, alkaline earths, carbonates, and metal salts of lead and silver.

Storage : It should be stored in air-tight containers, protected from moisture because it is hygroscopic in nature.

Uses : It is used as expectorant, diuretic in lead poisoning and systemic acidifier in mercurals treatment of urinary infection.

Dose : 300 mg to 1 g as expectorant.

(H-2) POTASSIUM IODIDE

Chemical Formula : KI

Molecular Weight : 166

Synonyms : Kalii iodidum, Pot. iod.

Standards : It contains not less than 99% of KI calculated with reference to dried substance at 105°C.

Preparation : It is prepared by treating a hot aqueous potassium hydroxide solution with iodine, a mixture of potassium iodide and potassium iodate is formed,

$$6KOH + 3I_2 \longrightarrow 5KI + KIO_3 + 3H_2O$$

Physical Properties : It is an odourless, colourless crystals; soluble in water and glycerin.

Test for purity :

(a) *Cyanides* : Dissolve 0.5gm in warm water, add drop of ferrous sulphate solution and 0.5ml of sodium hydroxide solution and acidify with hydrochloric acid; no blue colour is produced.

(b) *Iodates* : Two drop of dilute sulphuric acid, add 0.5gm of sample in water and a drop of starch solution; no blue colour is produced.

Incompatibility : It is incompatible with salts of bismuth, iron, mercury, potassium chlorate, strychnine hydrochloride, quinine sulphate and other alkaloidal salts.

Storage : It should be stored in air tight containers, protected from light.

Uses : It is used in prophylaxis and treatment of goiter. It is also used in the pre-operative treatment of thyrotoxicosis, expectorant and in the preparation of iodised salts as potassium iodate.

(H-3) ANTIMONY POTASSIUM TARTRATE

Chemical formula : $C_4H_4KO_7Sb$

Molecular weight : 324.9

Synonyms : Antimony pot. tart.; Tarter ,ematic.

Preparation : It is prepared by mixing antimony trioxide (Sb_2O_3) with potassium acid tartrate in 5:6 ratio. The paste is made and kept aside for 24 hours. It is boiled with water and liquid is filtered while hot. The crystals are formed on cooling.

$$Sb_2O_3 + 2C_4H_4O_6KH \longrightarrow 2C_4H_4O_7SbK + H_2O$$

Physical properties : It is a white, colourless, crystalline; odourless; sweet in taste and soluble in water but insoluble in alcohol.

Incompatibility : It is incompatible with acids, alkalis, soaps, salts of heavy metals and tannins.

Storage : It is a efflorescence compound. It should be stored in air-tight containers.

Uses : It is used as emetic; in small doses, it acts as expectorant. It is given in the treatment of KALA AZAR (schistosomiasis).

(I) ANTIDOTES

A substance which produces disease or death or ill health (alteration in normal physiology) is called poison. The acute poisoning may be accidental, occupational, self-medication, suicidal or criminal.

Antidotes are the agents which produces opposite physiological action or complex or prevent the absorption of poison in body fluids.

The antidotes can be classified as per their mode of action.

1. *Chemical antidote* : This type of agent combines with poison and changes poisonous behaviour of the compounds e.g., sodium thiosulphate which combines with toxic cyanide ion to form non toxic thiocyanide.
2. *Physiological antidote* : This type of agent produces opposite effect to that of poison or reverse physiological effect to neutralize poison effect. e.g. : sodium nitrite converts Fe^{++} to Fe^{+++} in order to bind cyanide ions with Fe^{++} haemoglobin, Fe^{+++} Methaemoglobin.
3. *Mechanical antidotes* : These agents prevent absorption of poison in the body either by increasing the rate of elimination (through urine) or by producing emesis (vomiting) or adsorbing the poison on their own surface to form inactive complex. e.g. : Diuretics (increases formation of urine) are agents which increase rate of elimination of poison,

cupric sulphate is emeting drug which is useful in gastric lavage in poisoning and activated charcoal acts by adsorption of poisonous substance.

Cyanide Poisoning : Cyanide poisoning may occur by inhalation of hydrocyanic acid as fumigates or other soluble inorganic seeds of peach, apricot, bitter almond, chokecberry or chemical used in photography and silver polishes. The cyanide poisoning symptoms are nausea, drowsiness, headache, dizziness, dyspanoea, convulsion and death.

SODIUM NITRITE

Chemical formula : $NaNO_2$

Molecular Weight : 69.00

Standards : Sodium nitrite contains not less than 97% and not more than 101% with reference to the substance dried over silica gel for four hours.

Preparations

1. It can be prepared by refluxing sodium nitrate with metallic lead so that sodium nitrate is reduced to sodium nitrite and forms water soluble sodium nitrite and water insoluble lead oxide.

$$NaNO_3 + Pb \longrightarrow NaNO_2 + PbO$$

2. When nitric oxide (NO) and oxygen (O_2) are passes into a solution of sodium carbonate (Na_2CO_3), it forms sodium nitrite and carbon dioxide.

$$4NO + O_2 + 2Na_2CO_3 \longrightarrow 4NaNO_2 + 2CO_2$$

Physical Properties : It is a deliquescent, colourless or slightly yellow crystal; soluble in 1 parts of water and 160 parts of alcohol.

Chemical Properties

1. It reacts with dilute acids to form nitrous acid which liberates nitrogen dioxide.

$$NaNO_2 + HCl \longrightarrow NaCl + HNO_2$$
$$3HNO_2 \longrightarrow HNO_3 + 2NO + H_2O$$
$$2NO_2 + O_2 \longrightarrow 2NO_2$$

2. When reacting with potassium salt or with stannous chloride in acidic medium, it acts as oxidising agent.

$$2KI + H_2SO_4 + 2HNO_2 \longrightarrow K_2SO_4 + 2NO + 2H_2O + I_2$$
$$2NaNO_2 + 4HCl + SnCl_2 \longrightarrow SnCl_4 + 2NO + 2NaCl + 2H_2O$$

3. Act as reducing agent : It reduces acidic pot. permanganate

$$2KMnO_4 + 3H_2SO_4 + 5NaNO_2 \longrightarrow K_2SO_4 + 2MnSO_4 + 3H_2O + 5NaNO_3$$

4. When heated with ammonium salt it liberates nitrogen.

$$NaNO_2 + NH_4Cl \longrightarrow NH_4NO_2 + NaCl$$

$$NH_4\,NO_2 \longrightarrow N_2 + H_2O$$

5. When it is added in a mixture of cobalt nitrate and acetic acid, it forms sodium cobalt nitrite $Na_2\,[Co(NO_2)_6]$

Incompatibility : Sodium nitrite is incompatible with oxidising agents, phenazone and caffeine citrate.

Storage : It should be stored in air-tight containers protected from moisture.

Uses : Sodium nitrite produces methaemoglobinaemia and cyanide ions combine with it to produce cyanmethaemoglobin, thus protecting essential enzymes from cyanide ions. This complex slowly metabolises to produce thiocyanate in presence of sodium thiosulphate and is excreted.

Dose : 30 to 120 mg.

3

Official Compounds of Iron, Iodine and Calcium

IRON

Iron is an essential element in the synthesis of haemoglobin. It is present in blood and tissues. In blood it transports oxygen from lungs to various organs and bicarbonate from tissues and plasma to lungs. It helps in the oxidation reduction of normal metabolism in tissues. Oxidation occurs in mitochondria due to ferritin moeity present in it. Iron is also associated with myoglobin, catalase, ferri doxin cytochrome and electron transport system.

The main source of iron is green leafy vegetables, grains, cereals, meat, egg yolk, etc. The daily dietary requirement is 10 mg in male and 18 mg in female but these may increase during growth, pathological bleeding, mensturation, pregnancy and any disease where blood loss occurs.

The ferric ions are used for external application in precipitation of protein and act as astringent, while ferrous ions are easily absorbed in G.I.T., hence given for internal use. The ferrous ion is a good source of iron in iron deficiency or anaemia.

FERROUS FUMARATE

Chemical formula :

$$\left[\begin{array}{c} HCCOO^- \\ \| \\ {}^-OOCCH \end{array}\right] Fe^{2+}$$

Molecular Weight : 169.91

Physical Properties : It occurs as reddish orange to reddish brown fine powder; slight odour; taste is astringent, slightly soluble in water and very slightly soluble in alcohol.

Storage : It is stored in well-closed containers.

Uses : It is used as source of iron (haematinics).

Dose : Initially 400-600 mg daily in divided dose and later 200 mg daily as maintenance dose.

FERROUS GLUCONATE

Chemical Formula : $C_{12}H_{22}FeO_{14}, 2H_2O$ or

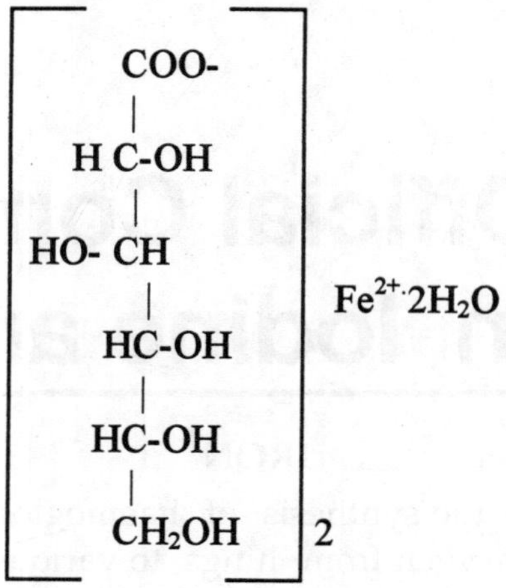

Molecular Weight : 482.17

Physical Properties : It is a yellowish grey fine granular powder; slightly burnt sugar like odour, fairly soluble in water and almost insoluble in alcohol.

Storage : It should be stored in air tight containers, protected from light.

Uses : It is also used as source of iron and in various anaemias.

Dose : The initial dose is 1.2 to 1.8 gm daily in divided dosage and 600 mg is maintenance dose.

FERROUS SULPHATE

Chemical Formula : $FeSO_4 \bullet 7H_2O$

Molecular Weight : 278.00

Synonyms : Green Vitriol.

Standards : It contains not less than 98% and not more than 105% $FeSO_4 \bullet 7H_2O$.

Preparations : When iron is treated with dilute sulphuric acid, iron dissolves to form ferrous sulphate and hydrogen gas is liberated.

$$Fe + H_2SO_4 \longrightarrow FeSO_4 + H_2$$

Physical Properties : It is a transparent, pale bluish green crystalline powder; odourless;

metallic astringent taste; freely soluble in water, very soluble in boiling water but practically insoluble in alcohol. This on exposure to moist air oxidises and becomes brownish yellow basic ferrous sulphate.

Chemical Properties

1. When it is treated with $KMnO_4$ in presence of H_2SO_4, it forms ferric sulphate.

$$2KMnO_4 + 3H_2SO_4 \longrightarrow K_2SO_4 + 2MnSO_4 + 3H_2O + 5O$$
$$(2FeSO_4 + H_2SO_4 + O \longrightarrow Fe_2(SO_4)_3 + H_2O) \times 5$$

$$2KMnO_4 + 8H_2SO_4 + 10\,FeSO_4 \longrightarrow K_2SO_4 + 2\,MnSO_4 + 5\,Fe_2(SO_4)_3 + 8\,H_2O.$$

2. When it is treated with ceric ammonium sulphate in acidic medium, it reduces ceric ion.

$$\underset{\text{(divalent)}}{2Fe^{++}} + \underset{\text{(tetravalent)}}{Ce^{++++}} \longrightarrow \underset{\text{(trivalent)}}{2Fe^{+++}} + \underset{\text{(divalent)}}{Ce^{++}}$$

Storage : Stored in tightly closed containers. It is slowly oxidised to ferric sulphate with atmospheric oxygen.

Incompatibility : It is incompatible with alkali, soluble carbonates, Ag and Au salt, potassium iodide, potassium and sodium tartarate, sodium borate, lead acetate, tannins etc.

Uses : It is used as haematinic in iron-deficiency. It is also used for dyeing fabrics and tanning of leather, disinfactant, manufacturing of ink etc.

Assay : It is assayed by direct redox titration. Weigh accurately about 1 gm and add 20 ml of dilute sulphuric acid and titrate with 0.1 N ceric ammonium sulphate using ferroin as indicator.

Factor : Each ml of 0.1 N ceric ammonium sulphate ≡ 0.0278 gm of $FeSO_4 \cdot 7H_2O$.

$$\underset{\text{(divalent)}}{\text{Ferrous salt}} + \underset{\text{(tetravalent)}}{\text{ceric salt}} \longrightarrow \underset{\text{(trivalent)}}{\text{Ferric salt}} + \underset{\text{(divalent)}}{\text{cerous salt}}$$

IRON-DEXTRAN INJECTION

When patient is not able to consume oral haematinic, then injection of iron dextran is given. It rises serum iron level rapidly in emergency due to surgery, pregnancy or premature iron deficient infant. It should not be given to the patients suffering from liver or kidney infections, allergic person, rheumatoid arthritis patients.

CALCIUM

Calcium is essential element for the tissue and for the blood. Blood contains approximately 2.5 millimoles/litre out of which about 7 mg is in ionized form and remaining is in colloidal form united with protein. The average daily requirement of calcium is 500 mg but larger amount is necessary during growth, pregnancy and lactation.

Inadequate calcium intake and absorption in children leads to rickets, which is due to malformation and imperfect calcification of bones and teeth; the demand by the infant to the mother during pregnancy and lactation may cause depletion of calcium from the maternal bones resulting in osteomalacia. Calcium balance is influenced by parathyroid hormone, excess of which mobilizes calcium from the bones, increases the calcium level and the proportion of ionic concentration as, compared with bound calcium level. Abnormally low level of calcium ion in the blood cause increased excitability of muscles and nerve tissues and may produce tetani or even convulsions, while high level of ionic calcium decreases muscular and nervous excitability. A high poisoning, in which it cause the deposition of lead in bone in the form of an insoluble salt. Injections of calcium salt should not be given during digitalis therapy.

CALCIUM CHLORIDE

Chemical Formula : $CaCl_2 \cdot 2H_2O$

Molecular Weight : 147.02

Physical Properties : It is a white crystalline powder; odourless; freely soluble in water, alcohol and very soluble in boiling water.

Storage : It should be stored in air-tight containers, at about 25°C.

Incompatibility : It is incompatible with soluble carbonate, phosphates, sulphates, tartarates, cardiotonics (digitalis).

Uses : It is used as source of calcium. It should not be given by intramuscular, intravenous or subcutaneous routes. A 10 ml is given in the treatment of cardiac arrest.

CALCIUM GLUCONATE

Chemical Formula :

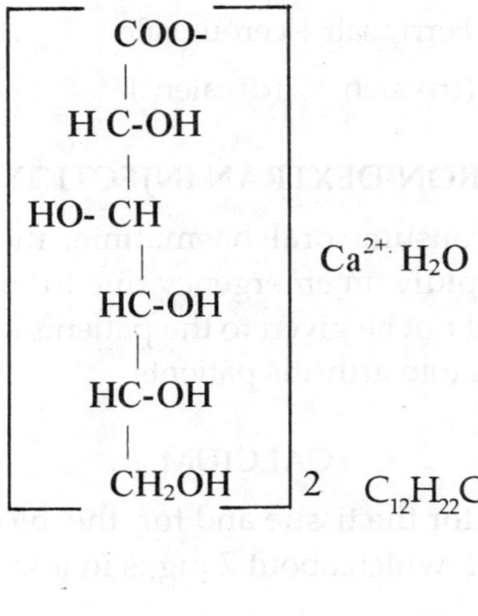

$C_{12}H_{22}CaO_{14} \cdot H_2O$

Molecular Weight : 448.4

Physical Properties : It is a white crystalline powder, odourless, tasteless, sparingly soluble in water, freely soluble in boiling water and insoluble in alcohol.

Storage : It should be stored in well-closed containers.

Incompatibility : It is incompatible with oxidising agent, phosphates and sulphates.

Uses : It is used as source of calcium and used in lead poisoning.

CALCIUM LACTATE

Chemical Formula : $(CH_3CHOHCOO^-)_2Ca$

Molecular Weight : 218.22

Physical Properties

It is a white granular powder, odourless, soluble in water and readily soluble in hot water. This is efflorescent powder, may contain pentahydrate.

Storage : It should be stored in air-tight containers.

Uses : It has similar use as calcium gluconate.

CALCIUM PANTOTHENATE

Chemical Formula :

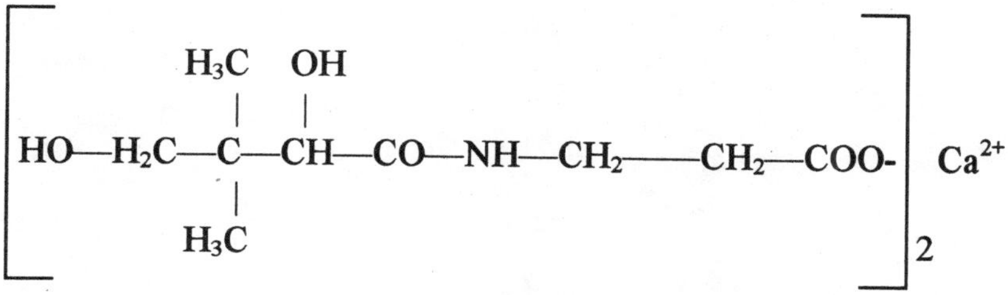

$C_{18}H_{32}CaN_2O_{10}$

Molecular Weight : 476.54

Physical Properties : It is a white, odourless, powder, bitter in taste, freely soluble in water, soluble in glycerin, practically in soluble in alcohol, chloroform and solvent ether.

Storage : It is hygroscopic in nature. It should be stored in well-closed containers.

Action and uses : This compound on hydrolysis gives calcium ion and pentothenate ion, which is an important water soluble vitamin B-complex moeity (Vitamin B_5). It serves as source of calcium and vitamin B_5.

IODINE

Iodine is an important trace element and externally it acts as anti-microbial agent, while its salts with sodium and potassium act as expectorant.

The weak iodine solution, strong iodine solution, aqueous iodine solution, iodine povidone solution and powders are potent astringent and anti-microbial agents on external application. It is a strong oxidising agent and oxidises proteins of micro-organisms. "Their compounds and preparations are well discussed in anti-microbial agents".

Internally iodine is an important trace element which is useful in converting thyronine to tri-iodo thyroxine (T_3) and tetra-iodo thyroxine (T_4). It is an important hormone secreted in thyroid gland. The T_3 and T_4 are active hormones secreted by thyroid gland. The deficiency of thyroid hormone is called "myxedema" or "hypothyrodism." It is characterized by slowing of metabolic process leads to bradycardia (decrease heart rate), decrease refluxes, weight-gain, constipation, drying of skin, puffiness of hands and face, poor memory, anxiety, nervousness, etc. In infants the development of physical and mental is impaired, results in "Cretinism" (Dwarfness and mental retardation). The lack of iodine leads to enlargement of thyroid gland, swelling at the neck, is called "goitre".

The excessive secretion of thyroid hormone is called "hyper-thyroidism". It leads to increase in basal metabolic rate (B.M.R). It is characterised by tachycardia (increase heart rate), anginal pain, cardiac arrhythmias, nervousness, anxiety and restlessness, sweating, weight loss. It may be due to disorder of immune response.

Therapeutically iodine is used in hyperthyroidism, fibrolytic agents in syphilis, leprosy, blastomycosis and actinomycosis, expectorant and bacteriocidal agent.

4

Major Intra and Extracellular Electrolytes

The body cells, and tissues require a constant osmatic balance environment with the help of body fluids, which contain inorganic and organic solutes. These are electrolytes, acids, bases, blood products, carbohydrates, amino acids and proteins. The body contains two types of fluids :

(i) Intracellular fluid - 40 to 50% of body weight.

(ii) Extracellular fluid - (a) Interstitial fluid - 12-15% of body weight.

(b) Plasma or vascular fluid - 4 to 5% of body weight.

These intra and extracellular fluids are separated from each other by semi-permeable membranes, which are permeable to certain ion such as Na^+, K^+, Mg^{++} Cl^-, HPO_4^{--} etc. and impermeable to macromolecules such as proteins, blood product, etc. The sodium chloride and calcium bicarbonate are major constituent of extracellular fluid while K^+, Mg^{++}, PO^{3-}_4 are major components of intracellular fluid.

The doses of individual ion are expressed in mEq/L. e.g., milliequivalent gram substance in one litre.

The equivalent weight is a mean of expressing combining power, obtained by dividing atomic or molecular weight by the valency.

$$mEq/L = \frac{\text{mg of substance}}{1 \div \text{Mol. wt./Eq/mole}}$$

It is derived from phosphoric acid represented as H_3PO_4, but more accurately as $PO(OH)_3$, also known as orthophosphoric acid.

Table 4.1
Inorganic ions, source, function & deficiency symptoms

S. No.	*Element*	*Amount in body*	*Food source*	*Functions*	*Deficiency*
1.	Sodium Na	1.8 gm/kg	Salt, milk, baking soda, salty food, etc.	1. Constituent in buffer 2. Acid base balance 3. Membrane Permeability 4. Muscle irritability 5. Maintain osmotic pressure	Caused by dehydration leads to acidosis, tissue atrophy
2.	Potassium K	2.6 gm/kg	Vegetables, fruits milk, legumes, meat	1. Constituent in buffer 2. Acid base balance 3. Membrane transport 4. Neuromuscular irritability	Acidosis leads to renal damage
3.	Calcium Ca	2.2 gm/kg	Milk, fish bones	1. Formation of bones and teeth 2. Clotting factor osteoporosis 3. Neuromuscular irritability 4. Membrane permeability	Child Rickets adult
4.	Magnesium Mg^{++}	0.5 gm/kg	Legumes, whole grains, nuts	1. It is a cofactor in phosphate transfering enzymes. 2. Constituent of bone and teeth	Muscular tremor, confusion, hyper irritability
5.	Iron Fe	75 mg/kg	Green leafy vegetables, egg yolk, liver,meat	1. Constituent of hemoglobin, ferredoxin, cytochromes 2. Cofactor in electron transport	Anemia.
6.	Cobalt Co	3 mg	Liver, Pancrease mushrooms	Constituent of Vitamin B_{12}	Pernicious anemia
7.	Manganese Mn	20 mg	Legumes, wheat germ, nuts, liver	Act as cofactor for enzymes like arginase, kinase	
8.	Chloride Cl^-	50 mEq/kg	Table salt	1. Electrolyte and osmotic balance 2. Acid-base balance	Hypochloremic alkalosis
9.	Sulphur SO_4	2-3 gm proteins	Plant and animal thiamine, B_4,etc.	Important constituent of muco polysaccharide,	Renal calculi
10.	Iodine (I^-)	100-140 Ugm	Seafoods and iodized salt	1. Regulate Cellular oxidation 2. Formation of thyroxin and t_3	Goiter, cretinism
11.	Fluoride (F)	1 PPM	Seafoods and drinking water	Fluoroapatite - tooth enamel.	Dental carries, mottled enamel
12.	Phosphate PO_4^{-3}	0.8 gm	Milk, egg, meat and legumes nut	1. Constituent of bone, tooth, buffer, ATP, NADP, FAD and Phosphoproteins	Osteomalacia, cardiac arrhythmia.

Electrolyte imbalance and replacement theory

The various conditions which cause disturbance in electrolyte balance and distrub, physiological systems such as internal homeostasis, fluid loss by elevated temp, vomiting;, blood loss, etc.The common conditions and their management by electrolytes during electrolyte imbalances are as under.

Major Physiological ion (ANIONS)

(i) *Chloride* : It is a major extracellular anion responsible for maintaining osmotic pressure, hydration, cation, anion balance in interstitial and vascular fluid. Food is the principal source of chloride ion. It is removed from the blood by glomerular filtration and possibly reabsorbed by renal tubules. Hypochloremia can be caused by

(a) Salt looseing during nephritis (associated with chronic pyelonephritis) which reduces reabsorption of chloride from renal tubules.

(b) Metabolic acidosis in diabetes mellitus and renal failure.

(c) Prolong vomiting with loss of Cl^- ion. Hypochloremic condition is seen in dehydration, decrease in renal blood flow during congestive heart failure, renal damage etc.

(ii) *Phosphate* : Phosphate (as HPO_4–) is the principal anion of intracellular fluid. Biochemically phosphate act as :

(a) Phosphate ester metabolises hexoses.

(b) Phosphoric acid anhydride linkage is the means of storage potential chemical energy, as adenosine triphosphate (ATP)

(c) It is an important constituent of physiological buffer system.

(d) It is essential for development of bone and teeth.

It is found in adenosine triphosphate (ATP) which releases 8000 calories per mole of energy by the hydrolysis of phosphate bond. Phosphate levels usually corelate with serum calcium value. The hyperphosphatemia may be due to hypervitaminosis D, renal failure caused by inability to excreate phosphate, and hypoparathyrodism. The basic aluminium carbonate is used to remove dietary phosphate by excreting it as aluminium phosphate.

(iii) *Bicarbonate* (HCO_3^-) : It is the second main anion present in extracellular fluid. It acts as an important physiological acid base buffer. Its deficiency causes metabolic acidosis and excess causes matabolic alkalosis.

(iv) *Sulphate* (SO_4^{2-}) : It is present in small quantity in the extracellular fluid. Its sources are animal and plant proteins having sulphur containing amino acid. It is important in detoxication mechanism and thio group containing organic compound in tissue respiration.

Major Physiological cations

(i) *Sodium (Na^+)* : It is a principal cation in the extracellular fluid responsible for maintaining normal hydration and osmotic pressure. The excess sodium is excreted by kidney. It is regulated by aldosterone which increases tubular reabsorption of sodium. Hyponatremia (low serum sodium level) are caused by: (i) Extreme urine loss in diabeties insipidus; (ii) Metabolic acidosis; (iii) Addison's diseases (decrease excretion of antidiuretic hormone, aldosterone); (iv) Diarrhoea and vomitting; (v) Kidney damage.

Hypernatremia (increase of serum sodium level) may be caused by: (i) Hyper adrenalism (cushing's syndrome) with increase aldosterone production; (ii) Severe dehydration; (iii) Brain injury and (iv) Excessive treatment with sodium salt.

If body is unable to eliminate sodium and its concentration increases, retention of water in the tissue due to osmotic balance causes edema.

(ii) *Potassium* (K^+) : It is a major intracellular cation present, as 23 times higher than extracellular compartment. The concentration difference is maintained by an active transport mechanism, called sodium potassium pump. During transmission of a nerve impulse, potassium comes out of the cell and sodium enters the cell.

Excess potassium salts serve as diuretic action, acidic condition, interference with sodium and potassium. Hypokalemia may change in myocardial function, flaccid and feable muscles, and low blood pressure. It can occur due to vomitting, diarrhoea, burns, haemorrhages, diabeties, over use of thiazides and alkalosis. It also alter electrocardio gram (ECG).

(iii) *Calcium* (Ca^{++}) : Calcium is found in extracellular fluid. It absorbs from upper part of intestine where content are acidic, or ionized water soluble acidic salt. Calcium is precipitated as dibasic phosphate, carbonate, oxalate, etc. The absorption is controlled by parathyroid hormone and cholecalciferol (Vitamin D_3).

(iv) *Magnesium* (Mg^{++}) : It is a main cationic intracellular fluid. It is an essential component of enzymes involving phosphate metabolism which requires ATP. An external injection produces depressant action and effects the cellular portion of neuron and neuromuscular junction similar to chloroform.

SODIUM CHLORIDE

Chemical Formula : NaCl

Molecular Weight : 58.44

Synonyms : Common salt, Natrii chloridum, Sod. chlor.

Standards : It contains not less than 99.5% and not more than 100.5% with reference to anhydrous sodium chloride.

Preparations

1. *From rock salt* : A brine is prepared with bore holes drilled in the rock salt strata and water is run. This brine is evaporated to obtained sodium chloride.
2. *From sea water/saltlake water* : The common salt is prepared by evaporation of salty water in shallow tanks or pans.

Physical properties : It is a white crystalline powder; saline taste; freely soluble in water and soluble in glycerin but slightly soluble in alcohol.

Test for purity

1. *For calcium and magnesium* : Not more than 0.005% impurities calculated as calcium with the help of 0.05 M disodium EDTA, ammonia-ammonium chloride as buffer and eriochrome black T as indicator.
2. *For iodide or bromide* : 2 gm substance is digested with 25 ml of warm alcohol, cool and filter and evaporate to dryness. Dissolve the residue and add 1 ml chloroform, 5 drops of dilute chlorine solution. It should not acquire a violet, yellow or orange colour.

Storage : It should be stored in air-tight containers.

Uses : It is an essential part of extracellular fluid contains about 140 milli equivalence/litre of sodium in extracellular fluid. Its various preparations are useful as parentrals, few are listed below.

Table 4.2

S. No.	*Preparation*	*Use*	*Remark*
1.	Sodium chloride inj.	Fluid and electrolyte replenisher irrigation solution	Sterile in single dose.
2.	Compound sod. chloride inj. (Ringer solution)	Irrigation solution use only.	For external
3.	Sodium chloride and dextrose injection	Fluid nutrient, electrolyte replenisher	Stored in single dose
4.	Mannitol and sodium chloride injection	Electrolyte replenisher cum diuretic	Useful for diabetic patients.
5.	Fructose and sodium chloride inj.	Fluid nutrient and electrolyte replenisher	--
6.	Comp. sodium lactate inj.	Electrolyte replenisher and systemic alkalizer.	Contraindicated in alkalosis.
7.	Lactate ringer injection	--do--	--do--

POTASSIUM-CHLORIDE

Chemical Formula : KCl

Molecular Weight : 74.55

Synonym : Kalii chloridum.

Preparations : It is prepared by the neutralizing hydrochloric acid with potassium bicarbonate

$$2\,HCl + K_2CO_3 \longrightarrow 2\,KCl + CO_2 + H_2O$$

Physical properties : It is a white crystalline-powder; odourless; saline taste; soluble in water but practically insoluble in dehydrated alcohol and ether.

Storage : Stored in well closed containers.

Uses : It is present in higher concentration in the intracellular fluid. The potassium ion produces depression on heart, acute muscular-weakness, paraesthesia and cardiac-arrhythmias.

Physiological Acid-Base Balance : The electrolytes are used to maintain physiological pH of body (7.38-7.42), because most of the metabolic reactions occur in this region. The normal pH of few selected body fluid are

Blood	: 7.4-7.5	Dudoenal fluid	: 5.5-7.5
Gastric Juice	: 1.5-1.8	Pancreatic Juice	: 7.5-8.2
Saliva	: 6.4-7.4	Urine	: 4.5-8.0
Gall bladder bile	: 5.5-7.7		

The body has two major buffer systems

(i) *Bicarbonate/Carbonic acid buffer system :* The system is due to exchange of CO_2 in lungs, plasma and tissue for the oxygenation. Carbon dioxide reacts with water in presence of carbonic anhydrase enzyme at plasma level to form carbonic acid. The plasma proteins act as buffers with decrease in the pH value due to formation of carbonic acid. Oxyhaemoglobin of erythrocytes activates the bicarbonate anion diffuses out of the erythrocytes and chloride anion diffuses in the cell.

$$CO_2 + H_2O \xrightarrow[\text{anhydrase}]{\text{Carbonic}} H_2CO_3$$

$$H_2CO_3 + K^+ + HbO_2^- \longrightarrow K^+ + HCO_3^- + HHb + O_2$$

This is called chloride shift. The above reaction is reversed in lungs where haemoglobin oxidises to oxyhaemoglobin and bicarbonate ion expels carbon dioxide from body.

$$O_2 + HHb + K^+ + HCO_3^- \longrightarrow K^+ + HbO_2^- + H_2CO_3$$

$$H_2CO_3 \longrightarrow H_2O\,CO_2$$

(ii) *Phosphate buffer system* : It is also effective in maintaining pH. If HPO_4^{--} and $H_2PO_4^-$ are present in 4:1 ratio than pH of the system will be 7.4. In kidney, the pH of urine changes from 4.5 to 4.8, corresponding to HPO_4^{--}: $H_2PO_4^-$ ratio of 1:99 to 1:100.

The acid excretion through kidney in 24 hours is equivalent to 20 to 40 ml of 1.0 N HCl. The steps for excretion of acids are as follows :

(a) The sodium salt of organic and mineral acids are removed from plasma during glomerular filtration.

(b) In renal tubules sodium reacts with carbonic acid formed by carbonic anhydrases enzyme.

$$Na^{+} + H_2CO_3 \longrightarrow Na^{-} + HCO_3^{-} + H^{+}$$

(c) This sodium bicarbonate returns to plasma and sodium ion reacts with phosphate or lactate and excretion of phosphate ion takes place.

The phosphate ion is very important for the oxidation of carbohydrates to generate high energy compounds like ATP while carbon dioxide is formed as a metabolite in kreb's cycle. (Aerobic breakdown)

When elimination of carbon dioxide through lungs is reduced (efficiency of respiration decrease) concentration of carbonic acid increases in plasma which leads to respiratory acidosis. It may be due to congestive heart failure, pneumonia, barbiturate poisoning or narcotic poisoning. In hyperventilation of lungs (increase in oxygen level) bicarbonate level of plasma decreases. It may be due to fever, anoxia, salicylate poisoning etc. This condition is called respiratory alkalosis.

In renal failure, ketosis, severe dehydration causes decrease in bicarbonate level of renal excretion and is called metabolic acidosis. While more amount of bicarbonate excretion is called metabolic alkalosis. It may be due to ingestion of large volume of antacids or depression of respiration.

SODIUM ACETATE

Chemical Formula : $CH_3COONa \bullet 3H_2O$

Molecular Weight : 136.08

Synonyms : Sodium acetate, Natrii acetas

Preparations : It is prepared by neutralization of acetic acid with sodium carbonate or bicarbonate.

$$2\,CH_3COOH + Na_2CO_3 \longrightarrow 2\,CH_3COONa + H_2O + CO_2$$

$$CH_3COOH + NaHCO_3 \longrightarrow CH_3COONa + H_2O + CO_2$$

Physical properties : It is a white, granular crystalline powder; odourless; saline bitter in taste; very soluble in water and soluble in alcohol.

Storage : Stored in tightly closed container to prevent efflorescence .

Action and uses : It is effective buffer in metabolic acidosis due to acute cholera. It maintains blood pH and plasma bi-carbonates by forming sodium bicarbonate. It is a source of sodium for haemodialysis and intraperitoneal dialysis.

POTASSIUM ACETATE

Chemical Forumla : CH_3COOK.

Molecular Weight : 98.14

Preparations : It is prepared by neutralizing acetic acid with postassium carbonate and potassium bicarbonate, till effervescence of carbon dioxide ceases.

$$2\,CH_3COOH + K_2CO_3 \longrightarrow 2\,CH_3COOK + H_2O + CO_2$$
$$CH_3COOH + KHCO_3 \longrightarrow CH_3COOK + H_2O + CO_2$$

Physical properties : It is a white crystalline powder. It is deliquescent, soluble in water and alcohol.

Storage : It should be stored in air tight container because of deliquescent property.

Uses : It is an alkalizer and acts as antacid in stomach. It has similar action as that of sodium acetate but acts as source of potassium.

SODIUM BICARBONATE

Chemical Formula : $NaHCO_3$

Molecular Weight : 84.01

Synonyms : Bakeing soda, Mitha-soda.

Uses : Sodium bicarbonate acts as antacid. Its 1.4% solution contains 167 milli moles/litres of sodium and bicarbonate. It is available in injection form. This injection may contain 0.01% disodium EDTA as sequisting agent. It is sterilized by filtration or by autoclaving in open container and carbon dioxide gas is passed after cooling.

It is used parentrally for systemic acidosis and in sodium replacement therapy in dehydration. It is also used in the treatment of methyl alcohol poisoning. It converts methyl alcohol to formic acid.

SODIUM CITRATE

Chemical Formula : $C_6H_5Na_3O_7 \bullet 2\,H_2O$

or

$$\begin{array}{l} CH_2COONa \\ | \\ C(OH)COONa\ .2H_2O \\ | \\ CH_2COONa \end{array}$$

Molecular Weight : 294.1

Synonyms : Natric citras, Sod. cit.

Preparation : It is prepared by neutralizing citric acid with sodium bicarbonate and sodium carbonate is added until effervescence ceases.

$$H_3C_6H_5O_7 \bullet H_2O + 3\,NaHCO_3 \longrightarrow Na_3C_6H_5O_7 \bullet H_2O + 3\,CO_2 + 3\,H_2O$$

$$2H_3C_6H_5O_7 \bullet H_2O + 2\,Na(CO_3)_2 \longrightarrow 2Na_3C_6H_5O_7 \bullet H_2O + 3CO_2 + 3H_2O$$

Physical properties : It is a white crystalline powder; odourless; saline in taste, very soluble in boiling water, freely soluble in water, practically insoluble in alcohol and in solvent ether. It is deliquescent in nature.

Storage : It should be stored in tightly closed container because of deliquescent nature.

Uses : It is oxidised in the tissues and it increases alkali due to liberation of carbon dioxide. It makes urine less acidic without neutralizing gastric secretion and is less laxative.

A 2.5% to 3.8% solution prevents clotting of blood in vitro condition. Citrates also act as chelating cation in benedict solution and are used in ferrous sulphate syrup.

POTASSIUM CITRATE

Chemical Formula : $C_6H_5K_3O_7 \bullet H_2O$

or

CH_2COOK
|
$C(OH)COOK\ .\ H_2O$
|
CH_2COOK

Molecular Weight : 324.42

Synonyms : Kalii citras, Pot. cit., Tripotassium citrate.

Preparation : It is prepared by neutralising the citric acid with potassium bicarbonate or potassium carbonate evaporating with constant stirring, granulating or powdering as required.

$$H_3C_6H_5O_7 \bullet H_2O + 3\,KHCO_3 \longrightarrow K_3C_6H_5O_7 \bullet H_2O + 3\,CO_2 + 3\,H_2O$$

$$2H_3C_6H_5O_7 \bullet H_2O + 3K_2CO_3 \longrightarrow 2K_3C_6H_5O_7 \bullet H_2O + 3CO_2 + 3H_2O$$

Physical properties : It is a white crystalline powder; odourless; saline in taste; very soluble in water; soluble in glycerin and practically insoluble in alcohol. It is a hygroscopic substances.

Storage : It is stored in well closed container because of its hygroscopic nature.

Uses : It has similar action to that of sodium citrate. It also acts as expectorant, diaphoratic, mild laxative and mild diuretic.

SODIUM LACTATE INJECTION

Chemical Formula : $NaC_3H_5O_3$ or $CH_3CHOHCOONa$

or

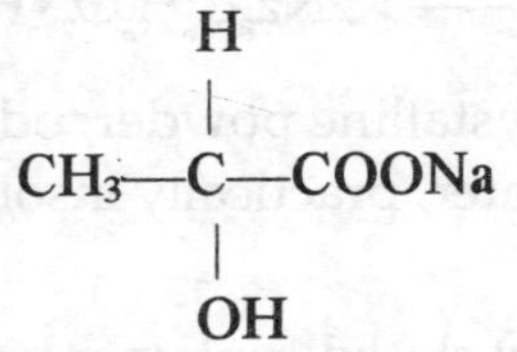

Molecular Weight : 112.04

Standards : It contains not less than 1.75% w/v and not more than 1.95% w/v of sodium lactate.

Preparations : It is prepared by neutralizing lactic acid or lactic anhydride with sodium carbonate or sodium hydroxide.

$$CH_3CHOHCOOH + NaOH \longrightarrow CH_3CHOHCOONa + H_2O$$

lactic acid

$$2\,CH_3CHOHCOOH + Na_2CO_3 \longrightarrow 2\,CH_3CHOHCOONa + H_2O + CO_2$$

lactic acid

Physical properties : It is a clear colourless liquid, miscible with water and alcohol. On storing for longer duration solid particles may appear; such solution should be rejected.

Action and uses : It is a systemic alkalizer, antiketogenic and is electrolyte replenisher fluid. In body tissue, it produces bicarbonate by the metabolic oxidation of lactates, thus it acts as metabolic alkalizer. It is generally given as systemic alkalizers in severe dehydration due to diarrhoea, diabeties mellitus and renal insufficiency.

Commercially sodium lactate injections contains 1.4% v/v solutions and compound sodium lactate injection (hartmann's solution for injection or ringer lactate solution for injection) contains 0.24% v/v solution.

AMMONIUM CHLORIDE INJECTION

Chemical Formula : NH_4Cl

Molecular Weight : 53.49

Synonyms : Sal ammoniac, Salmiac, Ammonium muriate.

Physical Properties : Ammonium chloride is a white crystalline powder; cooling saline taste; hygroscopic, freely soluble in water, glycerin but sparingly soluble in alcohol.

Uses : It act as expectorant, diuretic and acid base buffer in the body. It plays an important role in maintaining acid base equilibrium and reduces acidosis. The ammonia formed by the deamination of amino acids is excreted by kidney as ammonium ion and prevent sodium ion excretion, therefore, less amount of alkali (NaOH or Na_2CO_3) is excreted from body.

ORAL REHYDRATION SOLUTION (O.R.S.)

It is also called as combination of oral electrolytes. It is available as powders and solutions. These are required only after a heavy loss of water and electrolytes, may be due to severe vomiting, diarrhoea and prolonged fever. They contain essential electrolytes and a sufficient amount of water which can be easily absorbed. The typical concentration range are sodium 130 to 150 mEq/L, bicarbonate 28 to 55 mEq/L. (It may be substituted by lactate, acetate or gluconate in equivalent amounts), calcium 3 to 5 mEq/L and magnesium 3 mEq/L. These concentrations may vary as per the severity and loss of particular electrolyte. The commercial electrolyte combination therapy is done with oral rehydration salt (O.R.S). The natural O.R.S. therapy is done with butter milk, rice, starch water (kanji) with salt, coconut water, light tea, lemon sugar salt water, solution of cane sugar and salt (life saving solution), etc.

The W.H.O. recommended O.R.S. has following composition :

1. NaCl - 3.5 gm
2. $NaHCO_3$ - 2.5
3. KCl - 1.5 gm
4. Glucose - 20 gm
5. Water - 1 litre

5

Radio Pharmaceuticals

Some elements contain a few percentage of atoms which differ in atomic mass or atomic number as compared the parent atom. These are called isotopes. The production of radioactive substance involves bombardment of atomic nuclei with sub-atomic particles to produce unstable nuclei of some element. These unstable atoms have same chemical and physical properties as that of parent compound but differ in rate of chemical reaction (Kinetics).

Isotopes : Different kind of atoms of same element which have the same atomic number but different atomic masses. e.g., $^{12}.C_6$, $^{13}C_6$, $^{14}C_6$.

Isobar : The atoms of different elements which have the same mass number but different atomic number e.g., $^{40}K_{19}$, $^{40}Ca_{20}$.

Isotones : The atoms of different elements which have the same number of neutrons but different atomic number e.g., $^{14}C_6$, $^{15}N_7$

Half life : It is time required to disintegrate to half of its active original isotopes.

$$A = A_o e^{-\lambda t}$$

at half life when $t = t_{1/2}$ and

$$A = A_o e^{-\lambda/2}$$

$$t_{1/2} = \frac{0.693}{\lambda}$$

Where : A_o = Original active isotope

A = Active isotopes present at time

λ = Rate of dacay

Rations : There are two types of radioisotopes:

1. *Natural radionuclides* : These substances are having hi atomic weight (above 80), and their proton neutron ratio is varying from zero. Thes abstances produce radiations without stimulation.
2. *Artificial radionuclides* : These nuclides are originally inert to radiation but become active when controlled radiations are bombarded on it.

The radiation observed from radioactive substance are of three types: (i) α Particles (ii) β Particles (iii) γ Rays. This libration is continuous process and atoms try to give out energy and gets converted into a more stable nucleus.

1. **α Particles** : These particles are identical to Helium atom containing two protons and two neutrons. It's atomic mass is 4 and atomic number is 2(He_2^4). These are heavy particles with least penetrating power and more ionisation capacity.
2. **β Particles** : These are electrons emitted by nucleus having unit negative charge. If it is positively charged then it is called positron. These particles can penetrate in to the tissue up to 1 cm.
3. **γ Radiation** : These are electro-magnetic radiations similar to visible light. These radiations have a shorter wavelength but contains high energy with greater penetrating power as compared to visible light. These rays do not have any positive or negative charge.

Units of Radio-activity

Curie : This is equal to about one gram of radium which undergoes 3.7×10^{10} disintegration per sec.

Milli-Curie (mc) : It is 10^{-3} of Curie or when one gram of radium undergoes 3.7×10^{-7} disintegration per second.

Micro-Curie (μC) : This is 10^{-6} of Curie or when one gram of radium undergoes 3.7×10^{-7} disintegration per second.

Becquerel (Bq) : One disintegration per second is called Becquerel.

Electron Volt (eV) : It is the kinetic energy acquired by electron which is accelerated through a potential difference of 1 volt.

Roentgen (R) : It is unit of radiation exposed.

$$1\ R = 2.58 \times 10^{-4}\ C\ Kg^{-1}\ \ (C = \text{Coulomb})$$

Rad : It is the unit of absorbed dose 1 Rad = 10^{-2} Kg^{-1}

Biological effect of radiation : Radiations have very dangerous effect on biological tissue depending upon ability of radiations to penetrate the tissue, energy of radiation, surface area exposed, type of tissue and the dose of radiation. The radiation ionization promotes a number of irreversible changes in living cells.

These chemical changes alter the local pH or serve to initiate free radical chain reactions and forming peroxides and other toxic compounds. These and other events can create a hostile environment for tissue cell, leading to necrosis and, ultimately, complete destruction of the tissue or organ. The free radical formed from water can also abstract radicals from other molecules, resulting in the production of a variety of potentially toxic species which can alter the DNA in cells and causes cross linking between certain amino acids and proteins. The constant multiplying tissue like bone marrow mucosa of the gut, foetus, gonads etc. are most affected by radiations, irradiation sickness like nausea and vomiting are most common side effects of radiations. These effects may lead to change in protein sequence of DNA, and subsequent generation also produces similar effects.

(I) **Chain initiation**

$$H_2O \xrightarrow{\text{Radiation particles}} H\bullet + HO\bullet$$

(II) **Chain propagation**

$$H_2O + H\bullet \longrightarrow H_2 + \bullet OH$$

$$H_2O + OH\bullet \longrightarrow H_2O_2 + H\bullet$$

(III) **Chain termination**

$$H\bullet + H\bullet = H_2$$

$$OH\bullet + OH\bullet = H_2O_2$$

$$H\bullet + OH\bullet = H_2O$$

Measurement of radiations : When a radio active atom decay,it emits a particle or electromagnetic radiation or both, depending upon the mode of decay of the nucleus. It is the radiation resulting from the decay of atoms which is detected by the radiation measuring equipment. Various types of radiation measurement or detection instruments have been designed for the detection of a particular type of radiation. If radiation is to be detected with efficiency, the type of radiation must be known and the detector is selected accordingly.

The properties of radiations are utilized in their detection and measurement, e.g. the scintillation effect is measured in scintillation counter; the ionising effect is measured in ionisation chambers, proportional counters and Geiger Muller counter and the photograpic effect in autoradiography. The radiation detector can be operated either in current mode or in pube mode.

Radiation detectors are divided into two categories depending on :

(A) Collection of ions

(B) Collection of photons.

(A) ***Detectors utilizing ion-collection :***

1. Gas filled-detectors

   i. Electroscopes

   ii. Ion-chambers

iii. Proportional-counters

iv. Geiger-muller counters

2. Solid state detectors (semiconductor detectors)

i. Barrier-layer detectors

ii. Lithium-drifted detectors

(B) ***Detectors using photon-collection*** : These detectors are also called scintillation detectors.

i. Inorganic scintillators

ii. Organic scintillator

iii. Photogarphic-emulsions (Auto-radiography)

iv. Cerenkov detectors

v. Thermo-luminescence dosimeters.

vi. Track etch detectors

(A.1) Gas filled detectors

i. ***Ion chambers*** : Ionisation chambers are the simplest of all gas filled detectors. They depends upon the collection of all charges created by direct ionisation within the gas through the application of an electric field. The filled gas may be Helium (He), Argon (Ar), air etc. Gas filled detectors are available in various sizes and shapes. It consists of gas filled chambers filled with two electrodes kept at different electric potential and a measuring device which indicate the flow of electric current. Gas filled detectors are used as radiation monitors, dosimeters etc. They are operated in current mode.

ii. ***Geiger-Muller Counters*** *(G.M.counters)* : It is efficient for the detection of beta particles. It consists of a stainless steel cylinder (glass, silvered on the inner surface) which serves not only as the body of the tube but as the cathode. A fine wire, mounted coaxially, is the anode. The special gas mixture is also filled between anode and cathode. If a potential is maintained between the electrodes and the positively charged ions of gas produced by this ionization process are attracted and collected by the anode and cathode respectively. Each particle of radiations cause a brief pulse current to flow. Each pulse, representing the passage of a particle through the tube, then is recorded by a device such as a scaler to accumulate and indicate the total number of pulses.

iii. **Proportional counters** : These counters are similar to the G.M.-counter but modified by changing the gas composition and the shape of one or both of the electrodes.

An advantage of the proportional counter is its ability to distinguish between alpha and beta particles. If a radio active source emitting both alpha and beta particles is placed in the chamber, and the voltage applied to the chamber slowly increased. At a certain voltage only alpha particles are detected but at higher voltage both alpha and beta particles are detected.

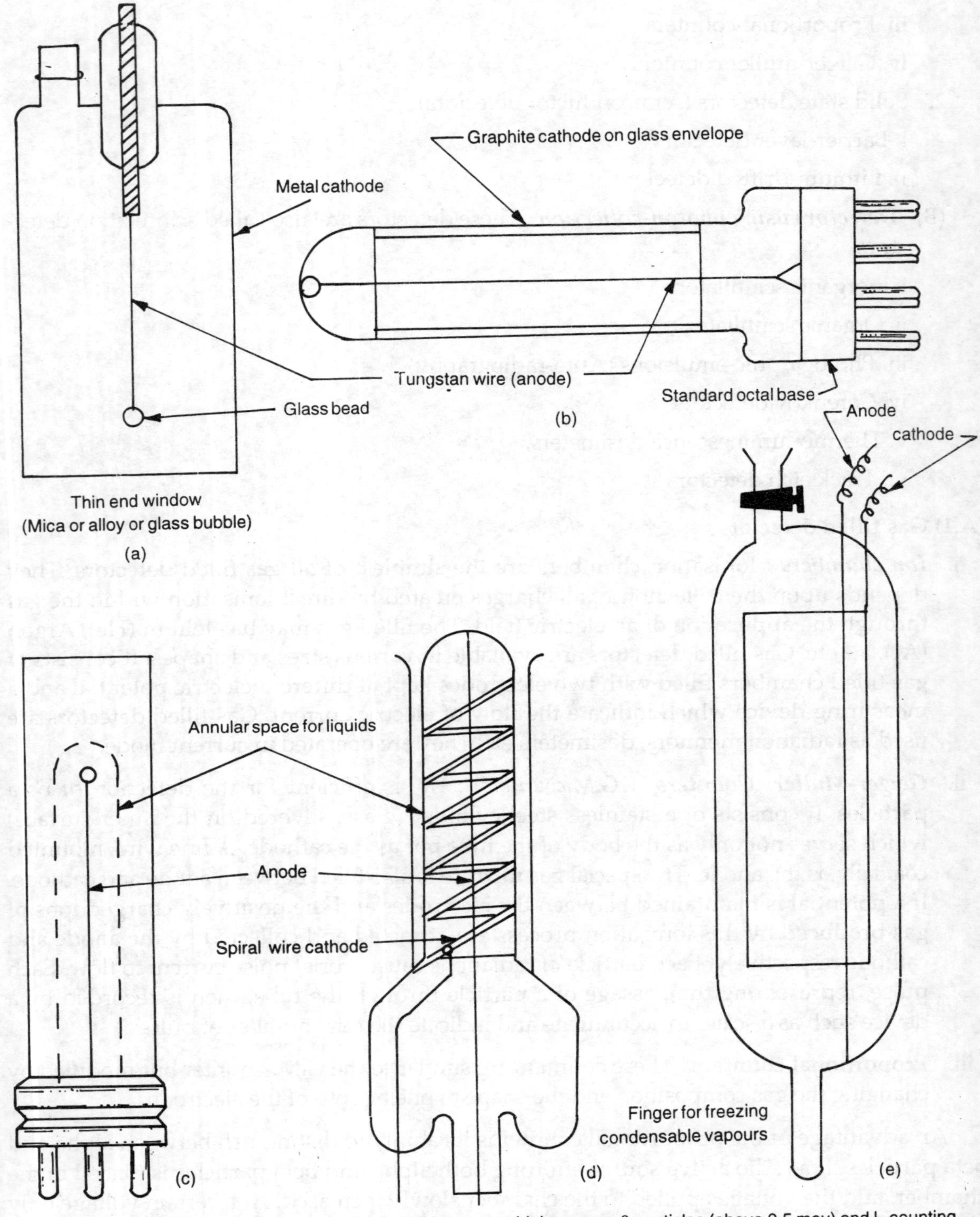

Geiger muller counters : (a) for solids, (b) for solids to count high energy β particles (above 0.5 mev) and L-counting, (c) & (d) for liquids, (e) for gases

(A.2) Semiconductor detectors : They are also called as solid state detectors. In this the band gap is very small of the order of 2 to 3 ev so the large number of electron hole pairs are formed per incident radiation quantum. This results in very good resolution of these detectors and they have the distinction of having the best energy resolution. Semiconductors may be of n-type (electron rich) and/or P-type (electron deficient). A semiconductor detector is diode of n and P semiconductors.

The absorption of incident-radiation leads to formation of e^- - h^+ pairs which move under the influence of applied electric field. The incident energy is proportional to the collection of electrons across a resistance lead to a voltage pulse.

Lithium drifted Ge, Ge(Li), Si (Li), Silicon-detectors and high purity Ge detectors (HP Ge) are most widely used in x-ray and gamma counting; surface barrier silicon detector is most commonly used for alpha spectroscopy and fission fragment kinetic energy spectra. Si(Li) detectors find wide application in conversion electron and x-ray measurements.

(B) Detectors using photo collection or scintillation-detectors : There are following scintillation detectors.

(B.1) Cerenkov detectors : These detectors are based on the light, which is emitted by fast charged particles through an optically transparent medium with refractive index more than one.

(B.2) Organic scintillators : The most widely used scintillators are anthracene having high-scintillation efficiency and stilbene having low scintillation efficiency. Organic scintillators are also used for simple and counting with 100% efficiency plastic scintillators, owing to their fast tinning response, are very useful is nuclear spectroscopy.

(B.3) Inorganic scintillators : Alkali halides are the most commonly used compounds. They are insulators and have a wide gap between the valency band and the conduction band. The activators are used to create excited states which decay by emission of the light in the visible-range. NaI(T1) is the widely used scintillator for gamma-counting due to its availability in desired shape and size and high detection efficiency. Other scintillators like CsI(Tl), LiI (Eu), BaF_2, CsI(Na), Bismuth germanate are used.

(B.4) Thermo luminescence dosimeters : Some of the inorganic-crystals in which electron-hole pairs have been formed due to radiation can trap the pairs and on heating lead to emission of light e.g. $CaSO_4$; MnLiF, CaF_2; Mn etc.

(B.5) Track-etch detectors : The principle involved in this is,when high linear energy transfer by ionising radiation, passing through a dielectric material create a trail of damaged molecules along their path. In some materials, the track can be visible upon etching in a strong acid or alkali solution. The damaged molecular tracks are etched faster than the bulk and look like pits on the surface. These tracks can be counted by viewing through a microscope. The commonly used track-etch materials are quartz, silica-glass, flint-glass, mica, cellulose nitrate, cellulose triacetate, polyethylene teriphthalate, etc.

(B.6) Photographic emulsions (Autoradiography) : Emulsion of halides of silver grains suspended in gelatin matrix and supported with cellulose acetate glass is principle constituent of ordinary photographic film. Silver halide grains are sensitive to ionising radiation which remain intact until they are developed. e.g., photographic emulsions are radiographic films and nuclear emulsion.

Uses : Photographic emulsions are very useful in determining and detecting, radiations in physiological studies of plants and animals.

HANDLING AND STORAGE OF RADIOACTIVE MATERIALS

A care should be taken to protect people and personnel from harmful radiation during handling and storage of radio active materials. These radiations may lead to serious complications such as blood cancer (leukemia). Lead (Pb) acts as shielding material for absorbing radiations. The following precautions are taken while working with radio detector, radioassays, tracer experiments, manufacturing or handling of radio active materials.

i. These materials should be handled with forceps or suitable instruments and direct contact should be avoided.

ii. Any substance which is taken internally (food, drinks, smokes, etc.) should not be carried in laboratory where radioactive materials are used.

iii. Sufficient shielding must be provided on protective clothes and other wears used by the person while handling the material.

iv. It should be stored in lead bricks shielded container with suitable level in a remote corner.

v. Radioactivity of the surrounding environment, where it is stored or handled should be monitored constantly.

vi. The final disposal of radioactive material should be done with great care to animals and environment.

The shielding effect can be achieved by thick concrete blocks or with pure distilled water layer. Water layer stops radioactive radiations and allows visible light to pass while concrete block of one sq meter thickness stops all the radiations.

Administration of Radioisotopes

The internal administration of Radioactive isotopes are orally or intravenously for diagnostic or therapeutic purposes. The treatment with Gamma radiation is given by focusing radiation directly on the area under treatment, this is called **teletherapy**.

The other mode of administration are implantion therapy, remote controlled shutter therapy, contact therapy etc. During the therapy certain areas are having high concentration, are called hot spot while low concentration areas are called spots. Sometimes isotopes are tagged with other molecules which releases isotope directly on the specific tissue. This prevents hazard of radio isotopes.

These therapeutic isotopes are used for their detrimental effect on tissue hence these isotopes should have sufficient energy to penetrate the tissue being treated but should not spread to surrounding tissues.

APPLICATION OF RADIOISOTOPES

Radioisotopes are useful in medicine, agriculture, pest control, age dating, sterilization, food preservation, etc. These agents are useful in varied field of pharmacy as therapeutic application, diagnostic agents, sterilization of parentral products and in research and development.

THERAPEUTIC APPLICATION

The radio isotopes are used as radiation source, either as external source or as internal source, not as tracers.

A. **External source** : These are used as sealed sources implanted in a tissue and dose is withdrawn by the removal of source from tissue. The dose should be calculated from effective half life of the isotopes and energy of radiation emitted. It is of three types

   I. *Teletherapy source* : The Co^{60} and Cs^{137} are neutron charged particles and emit radiation. It is used either as postoperative treatment of lesions or where surgical removal of lesions are not possible.

   II. *Surface source* : It is used for dermatologic and opthalmic applications. It contains emitters like P^{32} and Sr^{90}.

   III. *Extracorporeal irradation* : The x-ray generators are used initially on blood, followed by Co^{60} which leads to depletion of lymphocytes. It results in change in immunological response of individual.

B. **Internal source** : These type of agents are given as solutions, capsules, ampoules and injections. The following compounds are mainly used

1. *As infusions* : The Au^{198} and P^{32} are used as infusion. The gold Au^{198} is used in the treatment of peritoneal and pleural diffusions associated with malignant tumors in those cases in which fluid is accumulated in the abdomen or chest where ever large masses of tumor is not present.

The phosphate P^{32} is available as sodium phosphate. It is used in the treatment of polycythemia vera to decrease the rate of formation of erythrocytes. These are readily distributed to all tissues but concentrated in those tissues where proliferation is rapid.

2. *Intestinal implants* : The implants are spaced at intervals along a nylon ribbon. The Ir^{192} are used as intestinal implant therapy of tumors.

3. *Selectively absorbed* (Targeted therapy) : In this, P^{32}/I^{131}, and Y^{90} are main compounds which are well absorbed at the site of action or gets activated at the desired place.

The P^{32} and its salts are well absorbed in bone similar to normal phosphorus and become active in bone marrow and supress erythrogenesis. Yttrium Y^{30} has a strong affinity for chelating agents. The controlled chelation with N-hydroxy ethylenediamine triacetic acid can give systemic localization and selective irradiation of lymphatic structures. This is also used in leukemia and multiple myeloma.

Iodine I^{131} and I^{125} are useful as their sodium salt also. It is effective in hyperthyroidism by destroying thyroid tissue by radiation within the gland. It is very effective in the treatment of metastatic thyroid cancer. I^{131} labelled m-iodobenzyl guanidine can be given in metastases form in adrenal medullary carcinoma of brain, skull and axial skeleton.

C. **Others** : The human albumin microspheres of Tc^{99}, Pb^{203}, In^{111} and In^{113} are used for imaging lung coronary, urogenital, liver, gastro-intestinal, lymphatic and peripheral circulation.

Diagnostic application : The radioactive agents are useful for finding out various diseases, such as :

1. *Chromated albumin Cr^{51}* : It is used for detection and quantitation of gastro-intestinal protein loss and placenta localization. Cr(III) has a affinity for plasma proteins without affecting erythrocytes.
2. *Iodinated albumin inj* (I^{131}, I^{125}) : These are used to determine blood or plasma volumes, circulation times or cardiac output and for detection of brain tumors, in placental localization and in cisternography.
3. *Cyanocobalamin Co^{60}* : Its capsule and solutions are used for the absorption and deposition of vitamin B_{12} in megaloblastic anemias. The radio cyanocobalamine solution is useful in the diagnosis of pernicious-anaemia, a clinical condition characterised by deficiency or absence of this factor.
4. *Iron salts* : Ferric salts are used in lung imaging but ferrous citrate (Fe^{59}) is used for determination of metabolism of iron. It can also measure plasma iron, iron clearance half life, blood volume, haemoglobin formation and percentage of daily haemoglobin replacement.
5. *Fibrinogen I^{125}* : It is used to diagnose and localization of deep vein thrombosis, detection of renal transplant rejections and tumors.
6. *Gallium citrate* (Ga^{67}) : It is useful for diagnosis of lesions of the lung, breast, maxillary-sinuses and liver by imaging techniques. A positive uptake (Ga^{67}) is indicator of malignancies of lymphomas, bronchial carcinoma and Hodgkin's disease.
7. *Iodine I^{125} and I^{131}* : It is used for *in vitro* study of circulating insulin, either free or bound. It is used to study insulin kinetics and its plasma disappearance.
8. *Sterlization* : The cobalt-60 is used for the production of gamma rays which have high penetration power. This property is used to sterilize various thermolabile substances, syringes, needles, glove and clothes etc. used for parentral preparations or given parentrally.

9. *Miscellaneous* : The radio active material is used for research and development works, pest control, finding out steps of biological systems, carbon dating (age dating) and various other applications.

CONTRAST-MEDIA (RADIOPAQUES)

The radio active materials emits X-rays which can pass all the soft tissues of the body but are absorbed at hard tissues (bone). The rays produces a black spot on photographic plate by forming a complex with silver bromide. Wherever these rays does not form a bright spot of the similar shape, as that of X-ray absorbing object is formed.

Certain inorganic agents like Iodine or Barium salts can act as radio contrast media, by absorbing at soft tissues and these tissues becomes capable of absorbing X-rays. Thus provides a X-ray image of these tissues.

The barium is given only as barium salt which can not be ionized. The other barium salts get ionized and produces toxic effect of barium ion by stimulating all the muscles like G.I.T. muscles, CVS-muscles, respiratory muscles etc. The ionic iodine compounds are also toxic but their organic complex are covalently bonded (usually non-toxic), e.g., Iodinated aromatic organic acids and their derivatives, esters of fatty-acid from poppy seed oil etc.

The barium sulphate salt is given to identify location of ulcer in G.I.T. wherever ulcer is formed, broken mucosa retains barium sulphate, and this spot (ulcer) is identified with the help of x-ray film.

BARIUM SULPHATE

Chemical Formula : $BaSO_4$

Molecular Weight : 233.39

Synonyms : Barium-meal, Barii-sulfas.

Standardization : It contains not less than 97.5% and not more than 100.5% of $BaSO_4$.

Preparations : It is prepared from a solution of barium chloride with cold dilute H_2SO_4 or with soluble sodium sulphate. The barium sulphate is precipitated, washed and dried.

$$BaCl_2 + H_2SO_4 \longrightarrow BaSO_4 + 2\,HCl$$

$$BaCl_2 + Na_2SO_4 \longrightarrow BaSO_4 + 2\,NaCl$$

Physical Properties : It is a fine, white powder free from gritty particles; odourless; tasteless and practically insoluble in water, orgainc solvent and in dilute acid or alkali.

Chemical Properties : It dissolves in conc.sulphuric acid by forming bisulphate ion.

$$\underset{\text{Water insoluble}}{BaSO_4 + H_2SO_4} \longrightarrow \underset{\text{Water soluble}}{Ba(HSO_4)_2}$$

Storage : Stored in well-closed containers.

6

Identification Tests for Anions and Cations

ANIONS

I. Inorganic Anions

1. Preliminary test with dilute acids gives idea about some acid radicals. When mixture is treated with dilute H_2SO_4 or dilute HCl formation of volatile or gaseous product takes place. The presence of suspected radicals is confirmed by specific tests.

Table : 6.1

S. No.	*Anion*	*Observation*
1.	Carbonate/bicarbonate	Brisk effervesence, gas is colourless, odourless and turns lime water milky.
2.	Nitrate (NO_2^-)	Reddish brown fumes with pungent smell.
3.	Sulphide (S^{2-})	Colourless gas with smell of rotten egg is evolved.
4.	Sulphite $(SO_3)^{2-}$	Colourless gas with smell like burning sulphur
5.	Thiosulphate $(S_2O_3)^{2-}$	White turbidity
6.	Acetate $(CH_3COO)^-$	Vinager like smell.

CARBONATE

(a) 0.1 gm of sample being examined is taken in test tube containing 2ml of water (or prescribed solution). 2ml of 2M acetic acid is added in test tube and it is closed immediately with stopper fitted glass bent tube. Heat gently and collect the gas in 5ml of 0.1M barium hydroxide; a white precipitate is formed. Precipitate is dissolved on addition of an excess of dilute Hydrochloric acid.

$$CaCO_3 + 2CH_3COOH \longrightarrow (CH_3COO)_2Ca + H_2O + CO_2$$
$$Ba(OH)_2 + CO_2 \longrightarrow BaCO_3 + H_2O$$
$$BaCO_3 + HCl \longrightarrow BaCl_2 + CO_2 + H_2O$$

(b) Solution of sample is treated with solution of magnesuim sulphate, a white precipitate is formed.

$$Na_2CO_3 + H_2O \longrightarrow 2Na^+ + 2OH^- + CO_2$$
$$MgSO_4 + OH^- \longrightarrow MgOH^+ + SO_4^{2-}$$
$$2MgOH^+ + CO_3^{2-} \longrightarrow \underset{\text{white ppt}}{Mg_2(OH)_2CO_3}$$

BICARBONATES

(a) Carbon dioxide is evolved on boiling sample solution.

$$2NaHCO_3 \longrightarrow Na_2CO_3 + H_2O + CO_2$$

(b) When sample solution is treated with magnesuim sulphate, no precipitate is formed (distinction from carbonates), on boiling,a white precipitate is formed.

(In cold) $NaHCO_3 + MgSO_4 \longrightarrow$ No reaction

(On boiling) $2NaHCO_3 \longrightarrow Na_2CO_3 + H_2O + CO_2$

$$Na_2CO_3 + H_2O \longrightarrow 2Na^+ + 2OH^- + CO_2$$
$$MgSO_4 + OH^- \longrightarrow MgOH^+ + SO_4^{-2}$$
$$2MgOH^+ + CO_3^{2-} \longrightarrow Mg_2(OH)_2CO_3$$

(c) 0.1 gm of sample being examined is taken in test tube and 2ml of 2N acetic acid is added. Test tube is closed immediately with stopper fitted with glass bent tube. Heat gently and collect the gas in 5ml of 0.1N barium hydroxide solution, white precipitate forms that is dissolve on addition of an excess of dil hydrochloric acid.

$$NaHCO_3 + CH_3COOH \longrightarrow CH_3COONa + H_2O + CO_2$$
$$Ba(OH)_2 + CO_2 \longrightarrow BaCO_3 + H_2O$$
$$BaCO_3 + 2HCl \longrightarrow BaCl_2 + CO_2 + H_2O$$

THIOSULPHATE

(a) 0.1 gm of sample being examined is dissolved in 5ml of water and 2ml of hydrochloric acid is added, white precipitate is formed which turns yellow. Sulphur dioxide gas is liberated.

$$Na_2S_2O_3 + 2HCl \longrightarrow 2NaCl + H_2S_2O_3$$
$$H_2S_2O_3 \longrightarrow \underset{\substack{\text{White ppt} \\ \text{turns yellow}}}{S} + SO_2 + H_2O$$

(b) 0.1 gm of sample being examined is dissolved in 5ml of water and 2ml of ferric chloride solution is added. Dark violet colour is produced which quickly disappears.

$$Na_2S_2O_3 + 2HCl \longrightarrow 2NaCl + H_2SO_3 + S$$

$$3H_2SO_3 + 2FeCl_3 \longrightarrow Fe_2(SO_3)_3 + 6HCl$$

(Violet colour)

$$Fe_2(SO_3)_3 \longrightarrow FeSO_3 + FeS_2O_6$$

(c) Thiosulphate solution decolorise iodine and decolorised solution does not give the reaction of sulphate

$$2Na_2S_2O_3 + I_2 \longrightarrow 2NaI + Na_2S_4O_6$$

(d) Thiosulphate solution decolorises bromine solution, the decolorised solution give the reactions of sulphates.

2. When sample is treated with conc. H_2SO_4, it gives characteristic observation. The preliminary observations are :

Table : 6.2

S.No.	*Anion*	*Observation*
1.	Chloride (Cl^-)	Colourless gas with pungent smell is evolved, white fumes are produced when a rod dipped in ammonium hydroxide solution is brought near gas.
2.	Bromide (Br^-)	Reddish brown fumes which intensify on addition of MnO_2 vapour, passes in water makes solution yellow.
3.	Iodide (I^-)	Violet pungent fumes evolved, vapours turns starch paper blue.
4.	Nitrate (NO_3)$^-$	Light brown fumes evolved.
5.	Oxalate (C_2O_4-)$^-$	Colourless gas turns lime water milky and burns with blue flame.
6.	Fluride F$^-$	Colourless pungent smelling fumes and test-fube becomes greasy.
7.	Tartrate	On heating the mixture with conc. H_2SO_4, charring occurs and smell of burnt sugar is obtained.

BROMIDE

When sample is treated with conc. H_2SO_4,liberation of reddish brown fumes (intensify on addition of MnO_2) shows the presence of bromide. It is confirmed by following tests.

In testing for bromides in the presence of iodides, all iodine must be removed by boiling the aqueous solution with excess of lead dioxide.

(a) Dissolve a quantity of sample being examined equivalent to about 3mg of bromide ion in 2ml water or use 2ml of prescribed solution. Acidify with 2M nitric acid, add 1ml of 0.1M silver nitrate, shake and allow to stand. A curdy pale yellow precipitate is formed. Centrifuge, and wash the ppt. with 1ml of water three times. Precipitate is to be suspended in 2ml of water and 1.5ml of 10M ammonia is added to dissolve the precipitate .

$$NaBr + AgNO_3 \longrightarrow AgBr + NaNO_3$$

yellow ppt

(b) 10 mg of substance being examined is dissolved in 2ml of water and 1ml of chlorine solution. Bromine gas is evolved, which is soluble in 2 or 3 drops of $CHCl_3$, reddish solution forms. Add phenol solution, a white precipitate is produced.

$$2KBr + Cl_2 \longrightarrow 2KCl + Br_2$$

Reddish Solution
(In Chloroform)

$$C_6H_5OH + 3Br_2 \longrightarrow 3HBr + C_6H_2Br_3OH$$

Tribromophenol

CHLORIDES

(a) Sample equivalent to 2mg of chloride ion is dissolved in 2ml of water or use 2ml of prescribed solution. Acidifying sample with dilute nitric acid, add 0.5 ml of silver nitrate solution, shake, curdy white precipitate is formed which is insoluble in nitric acid but soluble in dilute ammonia after washing with water. It reprecipitate after adding dilute nitric acid.

$$NaCl + AgNO_3 \longrightarrow AgCl + NaNO_3$$
$$AgCl + 2NH_3 \longrightarrow Ag(NH_3)_2Cl$$
$$Ag(NH_3)_2 + 2HNO_3 \longrightarrow AgCl + 2NH_4NO_3$$

(b) Sample equivalent to 10ml of chloride ion and 0.2 gm of potassium dichromate taken in test tube and 1ml of sulphuric acid is added. Filter paper strip moistened with 0.1 ml of diphenylcarbazide solution turns violet red when placed over test tube.

$$K_2Cr_2O_7 + 4NaCl + 6H_2SO_4 \longrightarrow 2\,CrO_2Cl_2 + 2KHSO_4 + 4NaHSO_4 + 3H_2O$$

Chromyl chloride

$$CrO_2Cl_2 + H_2O + \text{Diphenylcarbazide} \rightarrow \text{Violet red Colour}$$

IODIDES

(a) Sample equivalent to 4mg of iodide ion is dissolved in 2ml of water or use 2ml of prescribed solution after acidifying with dilute nitric acid, add 0.5ml of silver nitrate solution and shake. Curdy pale yellow precipitate is formed.

Centrifuge and wash the precipitate three time by 1ml of water. Suspend the precipitate in 2ml of water and add 1.5 ml of 10M ammonia, the precipitate does not dissolve.

$$NaI + AgNO_3 \xrightarrow{HNO_3} AgI + NaNO_3$$

Pale yellow ppt

(Insoluble in NH_3)

(b) 0.2ml of a solution of sample being examined equivalent of about 5mg of iodide ion per ml or 0.2ml of prescribed solution, add 0.5ml of 1M sulphuric acid, 0.15ml of potassium dichromate solution, 2ml of water and 2ml of chloroform. shake, chloroform layer becomes violet or violet red.

(c) Add 0.5 ml of mercuric chloride solution to 1ml of sample solution being examined containing equivalent of about 5mg of iodide ion. Dark red precipitate is formed which is slightly soluble in an excess of reagent and very soluble in an excess of potassium iodide solution

$$2KI + HgCl_2 \longrightarrow HgI_2 + 2KCl$$

Mercuric Iodide

$$HgI_2 + 2KI \longrightarrow K_2HgI_4$$

Potassium mercuric iodide

NITRATES

(a) 15mg of sample being examined is dissolved in 0.5ml of water. Add 1ml of sulphuric acid carefully, mix and cool. Incline the tube and carefully add, without mixing, 0.5ml of ferrous sulphate solution, a brown colour is produced at the interface of the two liquids.

$$NaNO_3 + H_2SO_4 \longrightarrow NaHSO_4 + HNO_3$$

$$6FeSO_4 + 2HNO_3 + 3H_2SO_4 \longrightarrow 3Fe_2(SO_4)_3 + 2NO + 4H_2O$$

$$FeSO_4 + NO \longrightarrow FeSO_4 . NO$$

(Brown ring)

(b) Add sample substance in mixture of 0.1ml nitrobenzene and 0.2ml of sulphuric acid. Allow to stand for 5 minutes and cool in ice bath, add 5ml of water slowly and 5ml of sodium hydroxide solution. Add 5ml of acetone, shake and allow to stand; upper layer show an intense violet colour.

3. Some anions are identified by their characteristic chemical reaction in solution e.g. sulphate.

SULPHATE

(a) Dissolve a 50mg amount of sample in dilute hydrochloric acid and add barium chloride solution. A white precipitate insoluble in conc. nitric acid is obtained .

$$Na_2SO_4 + BaCl_2 \longrightarrow BaSO_4 + 2NaCl$$

(b) Add 0.1 ml of iodine solution to the suspension obtained in above test, it remains yellow but soon decolorises by adding dropwise stannous chloride solution (distinction from iodates). Boil the mixture, no coloured precipitate is formed (distinction from selenates and tungstates)

(c) Add 2ml of lead acetate solution in sample solution (50mg in 5ml H_2O), a white precipitate, soluble in ammonium acetate solution and in sodium hydroxide solution, is produced.

$$K_2SO_4 + (CH_3COO)_2Pb \longrightarrow PbSO_4 + 2CH_3COOK$$

4. Interfering acid radicals

Oxalate, borate, fluorides, tartrate and phosphae are acknowledged as interfering acid radicals because of there interference in the regular course of analysis of basic radicals. Their salts are soluble in acidic medium and becomes insoluble in alkaline medium. In analysis of basic radicals, after IInd group solution is made alkaline by ammonium hydroxide to precipitate IIIrd group; interfering radicals are also precipitated along with the third group. This creates a problem in systematic analysis. Detection of these radicals are confirmed by following tests .

ACETATES

(a) When sample being examined is heated with equal amount of oxalic acid. Acid vapours with characteristic odour of acetic acid are liberated.

$$CH_3COONa + COOH\text{-}COOH \longrightarrow \underset{\text{Acetic acid}}{CH_3COOH} + Na_2CO_3$$

(b) 1gm of sample being examined is warmed with 1ml of sulphuric acid and 3ml of ethanol (95%), until odour of ethyl acetate is evolved.

$$CH_3COOK + H_2SO_4 + C_2H_5OH \longrightarrow CH_3COOC_2H_5 + K_2SO_4 + H_2O$$

(c) 30mg sample is dissolved in 3ml of water, 0.25 ml of lanthanum nitrate solution, 0.1ml of 0.1M iodine and 0.05 ml of dilute ammonia solution is added. Heat carefully, blue precipitate appears within few minutes.

CITRATES

(a) Add solution of calcium chloride to neutral solution of substance being examined. After boiling the solution, white precipitate appears which is soluble in 6M acetic acid.

$$2Na_3C_6H_5O_7 + 3CaCl_2 + 4H_2O \longrightarrow \underset{\text{white ppt}}{Ca_3(C_6H_5O_7)_2 .4H_2O} + 6NaCl$$

(b) Dissolve the sample equivalent to about 50mg of citric acid in 5ml of water. Add 0.5ml of sulphuric acid and 3ml of potassium permanganate solution warm until the colour of permanganate is discharged. Add 0.5ml of a 10% w/v solution of sodium nitroprusside in 1M sulphuric acid and 4gm of sulphamic acid, make alkaline with strong ammonia solution dropwise until all the sulphamic acid has dissolved. On addition of an excess of strong ammonia solution a violet colour, which turns violet blue.

PHOSPHATE

(a) Mix 5ml of neutral sample solution (pH=7) to 5ml silver nitrate solution, light yellow precipitate is formed, the colour remains unchanged on boiling but readily soluble in 10M ammonia and in dilute nitric acid.

$$Na_2HPO_4 + 3AgNO_3 \longrightarrow Ag_3PO_4 + HNO_3 + 2NaNO_3$$

$$Ag_3PO_4 \longrightarrow 3Ag^+ + PO_4^{3-}$$

$$Ag^+ + 2NH_3 = [Ag(NH_3)_2]^+$$

Silver complex

(b) 1ml of prescribed solution is mixed with 1ml of ammonical magnesuim sulphate solution, white crystalline precipitate is formed.

$$Na_2HPO_4 + MgSO_4 + NH_3 + 6H_2O \longrightarrow MgNH_4PO_4.6H_2O$$

White crystalling ppt

(c) Mix 2ml of prescribed solution with 2ml of dilute nitric acid and 4ml of ammonium molybdate solution, and warm the solution, a bright yellow precipitate is formed,

$$H_3PO_4 + 12(NH_4)_2MoO_4 + 13HNO_3 \longrightarrow (NH_4)_3PO_4.12MoO_3 + 2NaNO_3 + 21NH_4NO_3 + 12H_2O$$

TARTRATES

(a) Sample is warmed with sulphuric acid, charring occurs and carbon monoxide, is evolved when ignited in blue flame.

$$CH(OH)COOH\text{-}CH(OH)COOH + 2H_2SO_4 \longrightarrow 3CO + CO_2 + 2SO_2 + 5H_2O$$

(b) Dissolve 20 mg of sample being examined in 5ml of water or use 5ml of prescribed solution. Add 0.05 ml of a 1% w/v solution of ferrous sulphate and 0.05 ml of hydrogen peroxide solution, transient yellow colour is produced. After the colour is disappeared add 2M sodium hydroxide dropwise, intense blue colour is produced.

$$CH(OH)COOH\text{-}CH(OH)COOH + H_2O_2 \longrightarrow C(OH)COOH{=}{=}C(OH)COOH$$

Intense blue colour

(c) Heat 0.1 ml of a solution containing equivalent of 2 mg of tartaric acid or 0.1 ml of prescribed solution on a water bath for 5 to 10 minutes with 0.1 ml of 2% w/v resorcinol solution and 3ml of sulphuric acid. Dark blue colour is produced, colour changes to red on cooling.

REMOVAL OF INTERFERING RADICALS

1. BORATE AND FLUORIDE

These radicals are removed by common method. When II group is identified during analysis of cations, then first boil off H_2S gas, filter, filtrate is taken in porcelain dish and keep on boiling till a paste is obtained. Then add 1-2 ml conc. HCl to this paste and evaporate to almost dryness with stirring. Repeat the process, 3-4 times with the same volume of conc. HCl. This ensures their removal. Use the filtrate for IIIrd and higher group analysis (Note : In absence of IInd group use the original mixture for the removal)

Principle : Decomposition takes as follows.

$$NaF + HCl \longrightarrow NaCl + HF \text{ (Volatile)}$$

$$Ca_3(BO_3)_2 + 6HCl \longrightarrow 3CaCl_2 + 2H_3BO_3$$

2. PHOSPHATE

(a) Acetate Buffer - $FeCl_3$ Method.

In case the IInd group is present then boil off H_2S gas from IInd-group filtrate otherwise use the clear solution of mixture. Add few drops of conc. HNO_3 (to convert Fe^{2+} to Fe^{3+} if any) and heat for one minute. Add 1gm solid NH_4Cl and then add NH_4OH drop by drop with shaking, until either a slight permanent turbidity (due to phosphate and hydroxides) is just obtained as until just alkaline. Add 2-3 ml of CH_3COOH and 10 ml of approx. 3M solution (or 2-3 gm solid) CH_3COONH_4 and shake. If any ppt. is formed, filter. If there is any residue left, test for si^{3+}, Ca^{2+}, and Fe^{3+}. If filtrate is reddish (if sufficient Fe^{3+} is already present to precipitate all PO_4^{3+} ions) do not add $FeCl_3$ solution. If filtrate is not reddish add neutral or freshly prepared solution of $FeCl_3$ dropwise with shaking. It produces buff colour ppt. of $FePO_4$, continue the addition of $FeCl_3$ solution with shaking HCl a permanent deep reddish brown colouration due to ferric acetate persist. Dilute the solution by adding large excess of water and then boil it vigorously for 2-3 minutes to precipitate excess of Fe^{3+}. Filter rapidly while hot. Residue left may contain phosphate of Fe, Al, Cr and also Fe $(OH)_3$. Use it for analysis of Al^{3+}, Cr^{3+} and Mn^{2+}. Filtrate left should be boiled in a evaporating dish to 20 ml. Add 0.5gm solid NH_4Cl and then dil NH_4OH until it gives odour of ammonia. Filter, if necessary. Reject the residue and test is negative, filtrate is free from PO_4^{3-}. Use it for IV, V and VI groups analysis.

Principle : First, by usual way of analysis, third and all higher group phosphates are precipitated. Ammonium acetate and acetic acid is added to make medium buffer of constant pH which dissolve all the higher group phosphates leaving behind the third group phosphate. Since third group cations may not be in sufficient quantity to remove PO_4^{3-} as $FePO_4$.

Fe^{3+} and PO_4^{3-} ions react with acetate ion to give ferric acetate which on dilution and boiling precipitates as basic ferric acetate.

$$FeCl_3 + CH_3COONH_4 \longrightarrow (CH_3COO)_3Fe + 3NH_4Cl$$

On dilution and boiling

$$(CH_3COO)_3Fe + 2H_2O \longrightarrow Fe(OH)_2(CH_3COO) + 2CH_3COOH$$

brown ppt

(b) Zirconium Nitrate method

If IInd group is present boil off H_2S completely from IInd group filtrate or take fresh solution. Add a few drops of conc. HNO_3 and heat for 1-2 minutes., add about 0.5-1gm solid NH_4Cl and stir well until dissolved. Now start adding the aqueous solution of zirconium nitrate dropwise with shaking till no more ppt of zirconium phosphate forms. Avoid large excess of reagent otherwise turbid supernatant liquid will form, this turbidity cannot be removed by filtration, heat to boiling and filter.

If any residue left, reject it. To the filtrate add little solid NH_4Cl, heat and add excess of NH_4OH with shaking till smell of ammonia, boil and filter. Residue left should be used for analysis of cations of IIIrd group and filtrate should be used for analysis of IV, V, VI group.

3. TARTRATE AND OXALATE

Both are removed by common method. If IInd group is present, boil off H_2S gas from IInd group filtrate and keep on boiling till a paste is obtained. If IInd group is absent, use the original mixture. To this, add 1-2 ml conc. HNO_3 to the past taken in porcelain dish and evaporate to almost dryness with stirring. Repeat the same treatment with the same volume of conc. HNO_3 3-4 times. In the last mixture residue with dil HCl and filter if necessary. Use the filtrate for IIIrd and higher group analysis as usual.

Principle : With Conc. HNO_3, the oxalate and tartrate are converted into oxalic and tartaric acids respectively. These acids are easily decomposed in presence of conc. HNO_3 as follows:

$$Na_2C_2O_4 + 2HNO_3 \longrightarrow 2NaNO_3 + H_2C_2O_4$$

$$H_2C_2O_4 + 2HNO_3 \longrightarrow 2NO_2 + 2CO_2 + 2H_2O$$

(Oxalic acid)

$$H_2C_4H_4O_6 + 8HNO_3 \longrightarrow 8NO_2 + 3CO_2 + CO + 7H_2O$$

(Tartaric acid)

II. Organic Anions

BENZOATES

(a) Add 0.5 ml of ferric chloride test solution to 1ml of 10% w/v neutral solution of sample. Dull yellow precipitate, soluble in ether, is formed.

$$3C_6H_5COO^- + 2Fe^{3+} + 3H_2O \xrightarrow{\text{Neutral}} Fe(C_6H_5COO)_3.\ Fe(OH)_3$$

Dull yellow ppt

(b) Warm gently the mixture of 0.2 g of sample and 0.3 ml of sulphuric acid, white sublimate is deposited on the inner walls of the tube and charring does not occur.

$$2C_6H_5COONa + \text{Conc. } H_2SO_4 \xrightarrow{\text{Heat}} \text{White sublimate}$$

(c) Dissolve 0.5 gm of sample in 10 ml of water, or use 10ml of the prescribed solution and add 0.5 ml of hydrochloric acid. The precipitate obtained after crystallization and drying, it melts at about 122ºC.

$$C_6H_5COONa + HCl \longrightarrow C_6H_5COOH + NaCl$$

LACTATES

(a) Add 1ml of bromine water and 0.5 ml of 1M sulphuric acid to 5ml of sample solution being examined containing the equivalent of about 5 mg of lactic acid or 5ml of prescribed solution. Heat on water bath, stir occasionally, until colour disappears and add 4gm of ammonium sulphate, 0.2 ml of 10% w/v solution of sodium nitroprusside in 1M sulphuric acid dropwise without mixing, then add 1ml of strong ammonia solution without mixing. Allow to stand for 30 minutes. Dark green ring appears at the interface of the two liquids.

SALICYLATES

(a) When 1ml of 10% w/v neutral solution is added with 0.5 ml of ferric chloride test solution, violet colour is produced which persists after the addition of 0.1 ml of weak acetic acid.

$$3C_6H_4(OH)COONa + FeCl_3 \longrightarrow [C_6H_4(OH)COO]_3Fe + 3NaCl$$

Ferric salicylate

(Violet colour)

(b) 0.5gm of sample being examined is dissolved in 10ml of water or use 10ml of prescribed solution. Add 0.5ml of hydrochloric acid, precipitate obtained after recrystallization with hot water and dried, melts at 159ºC

$$C_6H_4(OH)COONa + HCl \longrightarrow C_6H_4(OH)COOH + NaCl$$

(c) 0.5 gm of sample being examined is dissolved in 10 ml of water or use 10ml of prescribed solution and add 2ml of bromine solution, a cream coloured precipitate is formed

$$(OH)COOHC_6H_4 \xrightarrow{Br_2} C_6H_4OBr_2$$

Cream-Coloured ppt

CATIONS

All cations are divided into seven different groups.

1. Preliminary test for group. I radicals is carried out by addition of dil HCl,

LEAD

(a) Dissolve 0.1 mg of the sample being examined in 1 ml of dilute acetic acid or use 1 ml of the prescribed solution. Add 2 ml of potassium chromate solution; a yellow precipitate forms that is insoluble in 2 ml of sodium hydroxide solution.

$$Pb^{2+} + CrO_4^{2-} \longrightarrow PbCrO_4$$

$$PbCl_2 + K_2CrO_4 \longrightarrow PbCrO_4 + KCl$$

Lead chromate(Yellow ppt.)

(b) Dissolve 50 mg of the substance being examined in 1ml of dilute acetic acid or use a 0.5 ml of the prescribed solution. Add 10 ml of water and 0.2 ml of 1M potassium iodide; a yellow precipitate forms. Heat to boiling for one or two minutes, and allow to cool; the precipitate is reformed as glistening, yellow plates.

$$Pb^{+} + 2I^{-} \longrightarrow PbI_2$$

$$Pb(NO_3)_2 + 2KI \longrightarrow PbI_2 + 2KNO_3$$

Yellow ppt.

MERCURY

(a) Place 0.05 to 0.1 ml of a solution of the substance being examined on well-scraped copper foil; a dark grey stain which becomes shiny on rubbing is produced. Heat the dried copper foil in a test-tube; the spot disappears.

(b) To a solution of the substance being examined add carefully potassium iodide solution; a red precipitate is produced which is soluble in an excess of the reagent (mercuric compounds) or a yellow precipitate is produced which may become green on standing (mercurous compounds).

(c) To the prescribed solution add 2M sodium hydroxide until strongly alkaline; a dense, yellow precipitate is produced (mercuric compounds).

(d) To a solution of the substance being examined, add 6M hydrochloric acid; a white precipitate is produced which is blackened by adding dilute ammonia solution (mercurous compounds).

SILVER

(a) Dissolve 10 mg of the substance being examined in 10 ml of water or use 10 ml of the prescribed solution. Add 0.3 ml of dilute hydrochloric acid; a curdy, white precipitate is formed that is soluble in dilute ammonia solution. Add potassium iodide solution; a yellow precipitate is formed.

$$Ag^+ + Cl^- \longrightarrow AgCl$$

$$AgNO_3 + NaCl \longrightarrow NaNO_3 + AgCl$$

curdy white ppt.

$$AgCl + 2NH_3 \longrightarrow Ag(NH_3)_2Cl \text{ (Soluble complex)}$$

$$Ag(NH_3)_2Cl + KI + 2H_2O \longrightarrow AgI + KCl + 2NH_4OH$$

yellow ppt.

(b) Dissolve 10 mg of the substance being examined in 10 ml of water or use 10 ml of the prescribed solution and add 2 ml of potassium chromate solution; a red precipitate is formed which is soluble in nitric acid.

$$2AgNO_3 + K_2CrO_4 \longrightarrow Ag_2CrO_4 + 2KNO_3$$

Brick-red ppt.

2. Presence of Group-II redical is detected by passing H_2S gas in presence of HCl, ppt. of Sb, As, Bi are formed test.

ARSENIC

Heat 5ml of the prescribed solution on a water bath with an equal volume of hypophosphorus reagent; a brown precipitate is formed.

BISMUTH

(a) To 0.5 g of the substance being examined, add 10 ml of 2N hydrochloric acid or use 10 ml of the prescribed solution. Heat to boiling for one minute, cool and filter if necessary. To 1 ml of the solution obtained, add 20 ml of water; a white or slightly yellow precipitate appears which on addition of 0.05 to 0.1 ml of sodium sulphide solution turns brown.

$$Bi^{3+} + 3HCl \longrightarrow BiCl_3 + 3H^+$$

$$BiCl_3 + 4H_2O \longrightarrow Bi(OH)_3 + H_3O^+ + 2Cl^-$$

$$Bi(OH)_2Cl \longrightarrow BiOCl + H_2O$$

Bismuth oxychloride (white ppt.)

$$2Bi^{3+} + 3S^{2-} \longrightarrow Bi_2S_3$$

Bismuth sulphide
(black ppt.)

(b) To about 50 mg of the substance being examined, add 10 ml of 2N nitric acid or use 10 ml of the prescribed solution. Heat to boiling for one minute, allow to cool and filter if necessary. To 5 ml of the solution obtained,add 2 ml of a 10 percent w/v solution of thiourea; an orange-yellow colour or an orange precipitate is produced. Add 4 ml of a 2.5 per cent w/v solution of sodium fluoride; the solution is not decolorised within thirty minutes.

$$Bi^{3+} + 3HNO_3 \longrightarrow Bi(NO_3)_3 + 3H^+$$

$$Bi(NO_3)_3 + NH_2CSNH_2 \longrightarrow \text{Orange yellow ppt.}$$

Thiourea

ANTIMONY

Dissolve with gentle heating about 10mg of the substance being examined in a solution of 0.5 g. of sodium potassium tartrate in 10ml of water, allow to cool; 2ml of this solution, or 2ml of the prescribed solution, to which sodium sulphide solution has been added dropwise, a reddish-orange precipitate which dissolve on adding dilute sodium hydroxide solution.

$$2Sb^{+3} + 3S^{2-} \longrightarrow Sb_2S_3$$

3. Presence of Group III radicals is detected by adding NH_4Cl and NH_4OH (in excess) to the salt solution

ALUMINIUM

(a) Dissolve about 20 mg of the substance being examined in 2 ml of water or use 2 ml of the prescribed solution, add about 0.5 ml of 2N hydrochloric acid and about 0.5 ml of thioacetamide reagent; no precipitate forms. Add dropwise 2N sodium hydroxide solution; a gelationus white precipitate appears that redissolves on addition of further sodium hydroxide solution. Gradually, add ammonium chloride solution; the gelationus white precipitate reappears.

$$Al^{3+} + 3OH^- \text{ (from NaOH)} \longrightarrow Al(OH)_3$$

$$Al(OH)_3 + OH^- \text{ (excess NaOH)} \longrightarrow [Al(OH)_4]^- \text{ (soluble)}$$

Tetrahydroxyl aluminate ion

(b) Dissolve about 20 mg of the substance being examined in 5 ml of water or use 5 ml of the prescribed solution, add five drops of ammonium acetate solution and five drops of a 0.1 percent w/v solution of mordant blue III; an intense purple colour is produced.

$$Al^{3+} + CH_3COONH_4 + \text{Mordant blue III} \longrightarrow \text{Intense purple colour}$$

(c) To a solution of the substance being examined in water, add dilute ammonia solution until a faint precipitate is produced, and then add 0.25 ml of a freshly prepared 0.05 per cent w/v solution of quinalizarin in a 1 per cent w/v solution of sodium hydroxide.

Heat to boiling, cool, and acidify with an excess of acetic acid, a reddish-violet colour **is** produced.

$$Al^{3+} + 3NH_3 + 3H_2O \longrightarrow Al(OH)_3 + 3NH_4^+$$

$$Al(OH)_3 + \text{Quinalizarin} + NaOH \xrightarrow{\text{Boil, cool}} CH_3COOH \text{ excess reddish white colour}$$

Quinalizarin (or 1,2,5,8-Tetrahydroxyanthraquinone) regent is represented as:

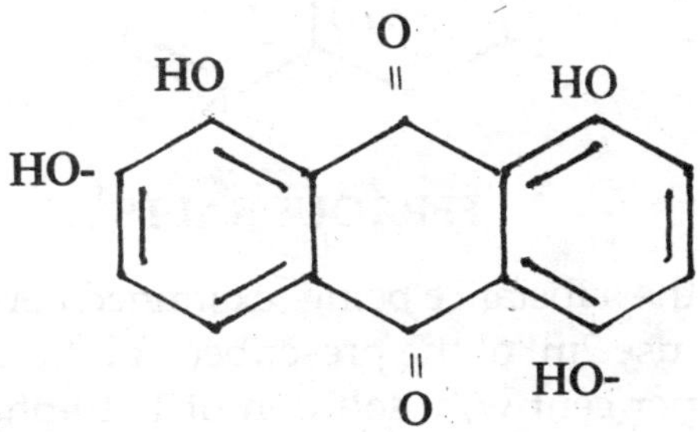

Quinalizarin

FERRIC SALTS

(a) Dissolve a quantity of the substance being examined equivalent to about 10 mg of iron in 1 ml of water or use 1 ml of the prescribed solution. Add 1 ml of a 5 per cent w/v solution of potassium ferrocyanide; an intense blue precipitate is formed which is insoluble in dilute hydrochloric acid.

$$4FeCl_3 + 3K_4Fe(CN)_6 \longrightarrow Fe_4[Fe(CN)_6]_3 + 12\,KCl$$

Pot. ferrocyanide Ferric ferrocyanide Prussian blue (Intense blue)

(b) To 3 ml of a solution containing about 0.1 mg of iron or to 3 ml of the prescribed solution, add 1 ml of 2N hydrochloric acid and 1 ml of ammonium thiocyanate solution; the solution becomes blood red in colour. Take two portions, each of 1 ml, of the mixture. To one portion, add 5 ml of solvent ether, shake and allow to stand; the ether layer is pink. To the other portion add 3 ml of 0.2M mercuric chloride; the red colour disappears.

$$FeCl_3 + 3NH_4SCN \longrightarrow Fe(SCN)_3 + 3NH_4Cl$$

Ferric thiocyanate(Blood red colour)

$$3HgCl_2 + 2Fe(SCN)_3 \longrightarrow 3Hg(SCN)_2 + 2FeCl_3$$

Mercuric thiocyanate

(Colourless)

(c) To 2 ml of a solution containing about 0.1 mg of iron or to 3 ml of the prescribed solution add acetic acid until the solution is strongly acidic and add 2 ml of a 0.2 per cent w/v solution of 8-hydroxy-7-iodoquinoline-5-sulphonic acid (Ferron Regent). A stable green colour is produced. The structure of ferron reagent is :

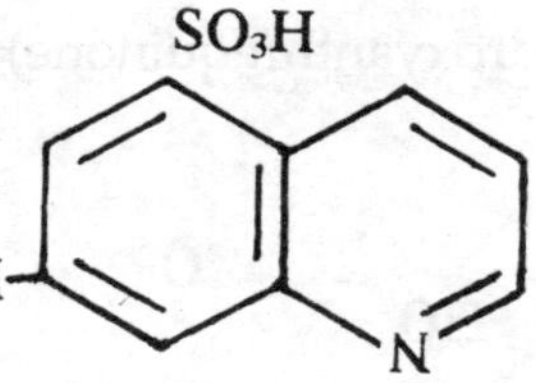

FERROUS SALTS

(a) Dissolve a quantity of the substance being examined corresponding to about 10 mg of iron in 2ml of water or use 2ml of the prescribed solution. Add 2ml of dilute sulphuric acid and 1ml of a 0.1 per cent w/v solution of 1, 10-phenanthroline. An intense red colour is produced; the colour is discharged by addition of a slight excess of 0.1N ceric ammonium sulphate. Red colouration is due to the formation of complex cation. The colour discharged by ceric ammonium sulphate due to oxidation of Fe^{2+} to Fe^{3+}.

(b) To 1 ml of a solution containing not less than 1mg of iron or 0.1ml of the prescribecd solution, add 1 ml of potassium ferricyanide solution; a dark blue precipitate is formed that is insoluble in dilute hydrochloric acid and is decomposed by sodium hydroxide solution.

$$3FeSO_4 + 2K_3[Fe(CN)_6]_2 \longrightarrow Fe_3[Fe(CN)_6]_2 + 3K_2SO_4$$

Pot. Ferricyanide Ferrous ferricyanide

(Dark blue) (Turnbull's Blue)

(c) To 1 ml of a solution containing not less than 1 mg of iron or to 1 ml of the prescribed solution, add 1 ml of potassium ferrocyanide solution; a white precipitate is formed which rapidly becomes blue and is insoluble in dilute hydrochloric acid.

$$2FeSO_4 + K_4Fe(CN)_6 \longrightarrow Fe_2[Fe(CN)_6]+2K_2SO_4$$

Ferrous ferrocyanide (White) air [O]

$Fe_4[Fe(CN)_6]_3$ (Prussian blue)

4. Presence of Group-IV radical is deteeted by passing H_2S gas in alkalne meduim (made by NH_4OH).

ZINC

(a) Dissolve 0.1 g of the substance being examined in 5 ml of water, or use 5 ml of the prescribed solution; add 0.2 ml of sodium hydroxide solution; a white precipitate forms. Add a further 2 ml of sodium hydroxide solution; the precipitate dissolve. Add 10 ml of ammonium chloride solution; the solution remains clear, but a flocculent, white precipicate forms on addition of 0.1 ml of sodium sulphate solution.

$$Zn^{+2} + 2OH^- \longrightarrow Zn(OH)_2 \downarrow$$

$$Zn(OH)_2 + 2OH^- \longrightarrow ZnO_2^{-2} + 2H_2O$$

$$ZnCl_2 + 2NaOH \longrightarrow Zn(OH)_2 \downarrow + 2NaCl$$

$$Zn(OH)_2 + 2NaOH \longrightarrow \underset{\text{Sod. zincate}}{Na_2ZnO_2} + 2H_2O$$

$$Na_2ZnO_2 + H_2S \longrightarrow \underset{\text{White ppt.}}{ZnS \downarrow} + 2NaOH$$

(b) Dissolve 0.1 g of the substance being examined in 15 ml of water, or use 5 ml of the prescribed solution, acidify with dilute sulphuric acid and add one drop of a 0.1 per cent w/v solution of copper sulphate and 2 ml of ammonium mercuri-thiocyanate solution; a violet precipitate is formed.

Copper salts alone do not form a precipitate with ammonium mercuri-thiocyanate. Zinc ion if present alone form white precipitate.

$$Zn + [Hg(SCN)_4]^{2-} \longrightarrow Zn[Hg(SCN)_4]$$

However, in presence of copper ion, the copper complex co-precipitates with zinc complex and the violet precipitate consists of mixed crystals of $Zn[Hg(SCN)_4] + Cu[Hg(SCN)_4]$.

(c) Dissolve 0.1 g of the substance being examined in 5 ml of water, or use 5 ml of the prescribed solution, add 2 ml of potassium ferrocyanide solution; a white precipitate is formed which is insoluble in dilute hydrochloric acid

$$2Zn^{+2} + [Fe(CN)_6]^{-4} \longrightarrow Zn[Fe(CN)_6] \downarrow$$

$$\underset{\text{Pot. ferrocyanide}}{2ZnSO_4 + K_4Fe(CN)_6} \longrightarrow \underset{\text{Zinc ferrocyanide (white ppt.)}}{Zn[Fe(CN)_6] \downarrow} + 2K_2SO_4$$

5. Presence of Group V radicals is detected by adding $(NH_4)_2CO_3$ in presence of NH_4Cl and NH_4OH.

BARIUM

(a) Barium salts impart a yellowish-green colour to a nonluminous flame, appearing blue when viewed through green glass.

(b) Dissolve about 20 mg of the substance being examined in 5 ml of dilute hydrochloric acid, add 2 ml of dilute sulphuric acid; a white precipitate forms which is insoluble in nitric acid.

$$Ba^{+2} + SO_4^{-2} \longrightarrow BaSO_4 \downarrow \text{(white ppt.)}$$

CALCIUM

(a) Dissolve 20 mg of the substance being examined in 5 ml of 5 M acetic acid or add 1 ml of glacial acetic acid to 5 ml of the prescribed solution. Add 0.5 ml of potassium ferrocyanide solution; the solution remains clear. Add about 50 mg of ammonium chloride; a white crystalline precipitate is formed.

$$Ca_2Fe(CN)_6 + NH_4Cl \longrightarrow \underset{\text{white ppt.}}{K_2Ca[Fe(CN)_6 \downarrow} + CaCl_2$$

(b) To a solution of the substance being examined, add a few drop of a solution of ammonium oxalate; a white precipitate is obtained that is only sparingly soluble in dilute acetic acid but insoluble in hydrochloric acid.

$$(NH_4)_2C_2O_4 \longrightarrow 2NH^+_4 + C_2O^{2-}_4$$

$$Ca^{+2} + C_2O^{2-}_4 \longrightarrow CaC_2O_4$$

$$CaC_2O_4 + 2HCl \longrightarrow CaCl_2 + 2\begin{array}{l}COOH\\ |\\ COOH\end{array}$$

Oxalic acid

(c) Dissolve 20 mg of the substance being examined in the minimum quantity of dilute hydrochloric acid and neutralise with dilute sodium hydroxide solution or use 5 ml of the prescribed solution; add 5 ml of ammonium carbonate solution; a white precipitate is formed which, after boiling and cooling the mixture, is only sparingly soluble in ammonium chloride solution

$$CaCl_2 + (NH_4)_2CO_3 \longrightarrow 2NH_4Cl + \underset{\text{white ppt.}}{CaCO_3 \downarrow}$$

6. Presevce of Group VI radical is detected by adding sodium hydrogen phosphate to alkaline solution of salt.

MAGNESIUM

(a) Disslove about 15 mg of the substance being examined in 2 ml of water or use 2 ml of the prescribed solution. Add 1 ml of dilute ammonia solution; a white precipitate forms that is redissolved by adding 1 ml of 2M ammonium chloride. Add 1 ml of 0.25M disodium hydrogen phosphate; a white crystalline precipitate forms.

$$Mg^{2+} + 2OH^- \underset{NH_4OH}{\overset{NH_3 \text{ solution}}{\rightleftharpoons}} Mg(OH)_2 \downarrow$$

The pressence of NH_4Cl prevents precipitation due to suppression of dissociation of NH_4OH due to common ion effect.

$$NH_4OH \longrightarrow NH_4^+ + OH^-$$

$$NH_4Cl \longrightarrow NH_4^+ + Cl^-$$

$$Mg^{2+} + NH_4^+ + PO_4^{-3} + 6H_2O \longrightarrow MgNH_4PO_4.6H_2O \downarrow$$

$$MgSO_4 + Na_2HPO_4 + NH_4OH + 6H_2O \longrightarrow \underset{\text{Crystalline white ppt.}}{MgNH_4PO_4.6H_2O \downarrow} + Na_2SO_4 + H_2O$$

(b) To 0.5 ml of a neutral or slightly acidic solution of the substance being examined add 0.2 ml of a 0.1 per cent w/v solution of titan yellow and 0.5 ml of 0.1N sodium hydroxide; a bright red turbidity develops which gradually settles to give a bright red precipitate.

7. Presence of VII or zero group is detected by testing individually radicals present in this group.

AMMONIUM SALTS

(a) Heat a few mg of the substance being examined with sodium hydroxide solution; ammonia is evolved, which is recognisable by its odour and by its action on moist red litmus paper.

$$NH_4Cl + NaOH \longrightarrow NaCl + H_2O + \underset{\text{Ammonia}}{NH_3 \uparrow}$$

(b) To the prescribed solution add 0.2 g of light magnesium oxide. Pass a current of air through the mixture and direct the gas that is evolved to just beneath the surface of 1 ml of 0.1N hydrochloric acid and 0.05 ml of methyl red solution; the colour of the solution changes to yellow. On addition of 1 ml of a freshly prepared 10 percent w/v solution of sodium cobaltnitrite, a yellow precipitate forms.

$$NH_4^+ + Na_3[Co(NO_2)_6] \longrightarrow (NH_4)_3 [Co(NO_2)_6] \downarrow$$

POTASSIUM

(a) Dissolve about 50 mg of the substance being examined in 1 ml of water or use 1 ml of the prescribed solution. Add 1 ml of dilute acetic acid and 1 ml of a freshly prepared 10 per cent w/v solution of sodium cobaltnitrite; a yellow or orange-yellow precipitate forms immediately.

$$3KCl + Na_3[Co(NO_2)_6] \longrightarrow K_3[Co(NO_2)_6 + 3NaCl$$

Potassium cobaltnitrite

(Orange-yellow ppt.)

(b) Dissolve 0.1 g of the substance being examined in 12 ml of water or use 2 ml of the prescribed solution. Heat the solution with 1 ml of sodium carbonate solution; no precipitate forms. Add 0.05 ml of sodium sulphide solution, no precipitate forms. Cool in ice water and add 2 ml of a 15 percent w/v solution of tartaric acid and allow to stand; a white crystalline precipitate forms.

$$KCl + H_2C_4H_4O_6 \longrightarrow KHC_4H_4O_6 + HCl$$

(c) Ignite a few mg of the substance being examined, cool and dissolve in the minimum quantity of water. To this solution add 1 ml of platinic chloride solution in the presence of 1 ml of hydrochloric acid; a yellow crystalline precipitate forms, which on ignition leaves a residue of potassium chloride and platinum.

$$PtCl_6^{2-} + 2K^+ \longrightarrow K_2[PtCl_6]$$

Pot. chloroplatinate

$$H_2PtCl_6 + 2KCl \longrightarrow K_2[PtCl_6] + 2HCl$$

Yellow ppt.

SODIUM

(a) Dissolve 0.1 g of the substance being examined in 2 ml of water or use 2 ml of the prescribed solution. Add 2 ml of a 15 per cent w/v solution of potassium carbonate and heat to boiling; no precipitate forms. Add 4 ml of freshly prepared potassium antimonate solution and heat to boiling. Allow to cool ice water and if necessary rub the inside of the test-tube with a glass rod; a dense, white precipitate is formed.

$$NaCl + KH_2SbO_4 \text{ ------> } NaH_2SbO_4\ v + KCl$$

Pot. antimonate Sod.pyroanti-

monate (white ppt.)

(b) Acidify a solution of the substance being examined with N acetic acid and a large excess of magnesium uranyl acetate solution; a yellow crystalline precipitate is formed.

$$NaCl + 3UO_2(CH_3COOH)_2 + Mg(CH_3COO)_2 + CH_3COOH \longrightarrow NaMg(UO_2)_3(CH_3COO)_9 + HCl$$

Greenish yellow ppt.

Definitions

Acid : A substance that ionizes in aqueous solution to yield a hydronium ion as one of the ions or a substance that donates proton or accepts electron. The strength of an acid depends upon the degree of ionization.

Acidimetry : The titration of free bases or those formed from the hydrolysis of salts of weak acids with a standard acid.

Addison's disease : Chronic adrenocortical insufficiency.

Alkalimetry : The titration of free acids or those formed by the hydrolysis of salts of weak bases with a standard alkali.

Amorphous : A substance having no definite shape or size, and which is coarse in nature.

Anaesthesia : A state chracterised by loss of sensation.

Angina : A severe constricting pain.

Anion : An ion or a radical having a negative charge.

Anorexia : Diminished appetite; aversion to food.

Antagonist : Something opposing or resisting the action of another.

Antiemetic : Preventing or arresting vomiting; a remedy that tends to control nausea and vomiting.

Arrhythmia : Irregularity; Loss of rhythm; denoting especially an irregularity of the heart beat.

Aseptic : Marked by or relating to asepsis.

Asthma : Paroxysmal attack of dificulty in breathing.

Atomic number : The number of protons in the nucleus of an atom.

Atomic weight : The weight of an atom with reference to the oxygen atom as 16.00.

Bactericide : An agent that destroys bacteria.

Bacteriostatic : Inhibiting or retarding the growth of bacteria.

Base : A substance that ionizes in a aqueous solution to yield hydroxide ion as one of the ion or as substance that possesses at least a lone pair of electrons or which accepts a proton. The strength of base depends upon the degree of ionization.

Bradycardia : Slowness of the heart beat.

Bronchitis : Inflammation of the mucous membrane of the bronchial tube.

Buffer solution : A solution which resists changes in hydrogen ion concentration, upon the addition of small amounts of acid or alkali.

Carcinoma : A malignant new growth derived from epithelial and glandular tissue; cancer.

Carminative : A agent that relieves flatulence.

Cation : An ion or radical having a positive charge.

Chelate : A complex of metal ion attached to atleast two co-ordinate bonds.

Choriocarcioma : A benign or malignant tumor of chronic tissue.

Chronic : Of long duration.

Colic : Spasmodic pains in the abdomen.

Complex ion : An ion which contains more than one element in a complex-combined form.

Convulsion : A violent spasm.

Decongestion : Relief of congestion.

Delequescence : A process in which a substance absorbs sufficient water to form a liquid.

Delirium : Abnormal mental condition based on hallucinations or illusions.

Dendrite : One of the thread like extension of the cytoplasm of a neuron.

Diarrhoea : An abnormally frequent discharge more or less fluid fecal matter from the bowel.

Diuresis : Extention of urine; commonly denotes production of unusually large volumes of urine.

Diuretic : Promoting the excretion of urine; an agent that increases the amount of urine.

Dysentery : A disease marked by frequent watery stools, often with blood and mucus.

Efflorescence : Loss of a water molecule from a crystalline substance.

Emesis : Vomiting; an act of vomiting.

Emetic : Bringing in or causing the act of vomiting; an agent that causes vomiting.

Enteric : Pertaining to the small intestine.

Equilibrium : A point at which the forward and reverse reaction rates are equal.

Equivalent Weight (of acid) : Is that weight of it which contains one atom of replaceable hydrogen ion. (For base) is that weight of it which contains one replaceable hydroxyl group i.e. 17.008g of ionizable hydroxyl. In redox agents, it is the molecular weigth divided by the number of electrons which 1 mol of substance gains or loses in the reaction.

Euphoria : A felling of well-being

Fugicide : Any substance that kills fungi.

Fungistatic : Having an inhibiting action upon the growth of fungi.

Gastric : Relating to the stomach.

Gastroenteritis : Inflammation of the mucous membrane of both stomach and intestine.

Germicide : An agent which is destructive to germs or microbes.

Glaucoma : A condition where the intraocular pressure is raised.

Glossitis : Inflammation of the tongue.

Gout : An inherited metabolic disorder occurring especially in men, charaterised by a raised but variable uric acid level, recurrent acute arthritis of sudden onset.

Haemorrhage : Bleeding a flow of blood especially if it is very profuse.

Hepatitis : Any inflammatory skin disease characterised by the formation of small vesicles in clusters caused by virus infection. When used alone, the term may refer to herpes simplex or to herpes zoster.

Hodgkin's disease : Progressive disease of reticulo-endothelial system shown in generalised enlargement of lymphatic glands and splenomegaly.

Hydration : Association of one or more molecular of water with an ion or molecular or compound.

Hypercalcemia : A reaction involving double decomposition in which water is one of the reactants.

Hygroscopic : A substance that has the ability to absorb water from the atmosphere or air.

Hyper-calcemia : Excess of calcium in the blood.

Hypertonic : A solution having an osmotic pressure higher than another solution, with which it is compared.

Hypotension : Abnormally low arterial blood pressure.

Hypotonic : A solution having an osmotic pressure lower than another solution, with which it is compared.

Inanimate : Not alive.

Indicator Paper : A paper impregnated with a solution of an indicator which shows a change of colour at an certain hydrogen ion concentration.

Inflammation : Multiplication of microorganisms in the body proper.

Infammation : The reaction of living tissue to injury, infection or irritation, characterised by pain, swelling, redness and breath.

Ionization potential : It is the energy required to convert a neutral atom from its lower most energy level to form a positive ion.

Isotonic : A solution having the same osmatic pressure as that of body fluids.

Keratitis : Inflammation of cornea.

Leukaemia : Progressive proliferation of abnormal leucocytes found in haemopoietic tissue, other organs, and usually the blood in increased number.

Ligands : A compound having lone pair of electrone and forms coordinate bound with metals.

Myeloma : A tumour composed of cell derived from haemopoietic tissue of the bone marrow.

Nephritic : Pertaining to the kidneys.

Molar solution (M) : A solution containing one mole (molecular weight in gm) per thousand ml of solution.

Neutralization : A chemical reaction in which acid and base react to form water and salt.

Normal solution (N) : A solution containing one gram equivalent weight per thousand ml of solution.

Oedema : Abnormal infiltration of tissue with fluid.

Osteomalacia : A disease characterised by a gradual softening and bending of the bones with more or less severe pain.

Osteroporosis : Loss of bone density due to excessive absorption of calcium and phosporous from the bone.

Osmosis : A process wherein certain constituents of a solution diffuse through a semipermeable memberane.

Oxidation : The process which results in the loss of one or more electrons by atoms or ions.

Oxidising agent : A substance which brings oxidation by accepting an electron.

pH : Negative log of hydogen ion concentration

Polymorph : A state wherein substance exists into one or more crystalline forms.

Prophylaxis : The prevention of a disease.

Pruritus : Itching.

Reduction : The process which results in the gain of one or more electron by atoms or ion.

Reducing agents : A substance which brings about reduction by donating an electron.

Rheumatism : A non-specific term embracing a diverse group of disease and syndromes which have in common, disorder disease of connective tissue and hence usually present with pain, or stiffiness, or swelling of muscles and joints.

Rubefacient : An agent which causes a reddening of the skin.

Scabies : An eruption due of *Sarcoptes scabies var. hominis.*

Sepsis : The presence of various pus-forming and other pathogenic organisms, or their toxins, in the blood or tissues.

Sinusitis : Inflammation of the lining membrance of any sinus, especially of one of of the prenasal sinuses.

Solute : A substance (the component of solution) which is dissolved in a solvent.

Solvent : A substance (the component of solution) which dissloved solute.

Styptic : Astringent; arresting haemorrhage by means of an astringent quality.

Techycardia : Rapid beating of the heart.

Tetany : A condition of muscular hyperexcitability in which mild stimuli produce cramps.

Ulcer : A lesion on the surface of the skin or a mucous surface.

Index